D0204782

The Challenge of the Atmosphere

Books by O. G. Sutton

The Challenge of the Atmosphere
Understanding Weather
Compendium of Mathematics and Physics
(with D. S. Meyler)
Mathematics in Action
Micrometeorology
The Science of Flight
Atmospheric Turbulence

THE CHALLENGE OF THE ATMOSPHERE

O. G. SUTTON

Harper Modern Science Series
edited by James R. Newman
Harper & Brothers, New York

THE CHALLENGE OF THE ATMOSPHERE

For information address
Harper & Brothers,
49 East 33rd Street, New York 16, N.Y.
Library of Congress catalog card number: 61–6442

FIRST EDITION

CONTENTS

PREFACE

This book is intended to give the reader a picture of the science of the atmosphere in its modern dress. It is written around the theme, familiar to meteorologists but possibly not to others, that climate and weather are manifestations of the activities of a complex of interlocking motion systems of varying size, driven by the energy of sunlight. The atmosphere presents to the meteorologist a challenge to produce an orderly account of this vast machine.

Although meteorology is now primarily a branch of mathematical physics, I have tried to write of these matters in terms that I hope can be understood by all. In this I have been greatly helped by Dr. James R. Newman, who made many useful suggestions when the book was in manuscript. I am also indebted to my wife for help in proofreading and to the publishers for their patience and skill in bringing the book into production.

O. G. Sutton

Sunninghill, Berks
England
June, 1961

PROLOGUE

Meteorology is a science; weather forecasting is a profession. Both are concerned with the atmosphere, the thin skin of air which clings to the Earth as it journeys through space. The atmosphere also moves over the surface of the Earth in complicated patterns in which, and in the accompanying weather, the meteorologist tries to trace sequences of cause and effect. The forecaster uses the knowledge so gained to predict the course of the weather. As yet neither the theoreticians nor the forecasters have entirely succeeded in their tasks. The scientist cannot always explain the movements of the atmosphere and forecasts sometimes mislead.

When compared with advances in other fields, progress in the science of the atmosphere seems painfully slow. It is not immediately obvious why this should be. Meteorology is a branch of physics and is therefore founded on mechanics, a set of rules propounded chiefly in the seventeenth and eighteenth centuries, by which the behavior of a physical system can be predicted whenever the configuration of the system, the forces which act upon it, and its state at some convenient zero of time can be specified. In mechanics the rules of forecasting are embodied in equations, and before the advent of high-speed computing machines prediction in physics usually required the discovery of the exact solution of the equations applicable to the problem in hand, preferably in the shape of a closed mathematical formula. Today we are content with approximate solutions tabulated by machines.

Some of the most spectacular and convincing achievements of classical mechanics have been in astronomy, and when events such as solar eclipses, once thought to be of supernatural origin, were not only explained but predicted infallibly years in advance by methods available to all it seemed that it could only be a matter of time before all natural processes were brought into similar relatively simple deterministic schemes. But with improved instruments and observations it became necessary to extend the mathematical analysis to natural systems far more complicated than that of the Sun and its planets, such as a swarm of molecules in a gas, whose configuration could not be specified in simple geomettrical terms and in which the forces at work could be deduced only from measurements of bulk properties, such as pressure and temperature. The Newtonian equations deal only with observables (that is, measurable quantities), but in the world of molecules the only observables are averages. This dilemma brought the notion of probability into physics. Even when the temperature of a gas has been measured with meticulous accuracy, all that has been learned is the level of the average energy of a vast concourse of molecules, and for an individual molecule we can never do more than to state the probability that its kinetic energy lies between certain limits. Statistical mechanics was invented not because Newtonian mechanics is false but for the more fundamental reason that it is too precise for the study of every kind of natural system, and as our measurements probe still further into the heart of matter the picture becomes still more blurred, until all that is left is the ghost of a particle, an entity of which we know little more than that its behavior conforms to numbers extracted from the solutions of certain equations.

There is a like situation in meteorology. In the study of the atmosphere we are compelled to employ a kind of statistical mechanics because it is impossible to take account of every cloud, every gust of wind, and every shower of rain. The original concept of air moving around the globe in well-defined streams, interspersed with regular vortices, has long been abandoned except by the makers of atlases. The real atmosphere is a system of almost unbelievable complexity, which differs from the solar system in that there is no one dominating force like gravity, but many, which add to the difficulties by forming innumerable feed-back loops. A forecast of weather can never be more than a statement of probabilities.

The analysis of atmospheric motion is a challenge to man's imagination and technical skill. As human beings we are aware of only a small part of the atmosphere at any time. It needs imagination to think of the hurricane which is battering at our houses as a dynamical system which formed in the placid tropical seas by a process which basically is no more than that involved in the condensation of water from the spout of a kettle. It needs skill to analyze this whirlpool of roaring winds and torrential rain as a thermodynamic process which follows the same laws of physics that are studied in the quiet of the laboratory. Yet this is being done every day with growing skill and deeper confidence in the outcome.

In this book an attempt is made to show meteorology as an analytic and deductive science, rather than a means of describing and cataloguing varieties of weather. The underlying theme is that there is in the atmosphere a hierarchy of motion systems, ranging from the great semipermanent rivers of air that flow around the globe to the tiny local eddies that are responsible for the processes on which life ultimately depends, such as the transfer of water from the surface to the air and back again as rain and snow. The challenge of the atmosphere is to extract some measure of rule from this turbulent system, to explain and predict its behavior, and perhaps someday to control it. Meteorology is the most difficult of all the earth sciences, and to its followers the most fascinating, because it is concerned with the least known part of our planet, its restless ocean of air.

The Challenge of the Atmosphere

CHAPTER 1

CLIMATE AND WEATHER

Meteorology, like all science, began in superstition and folklore. Man must have recognized the rhythm of the seasons and the effects of good and bad weather on his crops at a very early stage of civilization. At a later but still early stage he must have felt the urge to explain and, in particular, to assign causes to what are now recognized as natural extremes of weather. A disastrous flood, a drought, or a long severe winter arouses even today the desire to trace back the event to some specific cause. Our remote ancestors, at least, had few doubts. Bad weather indicated the wrath of the gods or was the work of evil spirits; good weather was a reward of virtue.

Mankind is not yet entirely free of this philosophy, but in science it is accepted that large variations in the physical state of the atmosphere are inevitable and arise, in a manner not yet completely understood, chiefly as a result of the particular composition of air and the fact that the atmosphere rests on a spinning body of irregular surface unequally heated by the Sun. The state of the atmosphere at any instant is referred to as its weather,[1] and it is commonplace that weather changes rapidly. If the time element is largely eliminated by taking averages of observations of weather over long periods, the result is called the *climate* of a region.

The Nature of Climate

In any discussion of climate it is important to recognize that what is experienced and observed is weather; climate is deduced

[1] In professional meteorology, and particularly in the coded messages by which meteorological information is passed around the world, the word "weather" is used in a more restricted sense, namely, to indicate the state of the sky and whether there is precipitation.

from weather and in this sense is a fiction, a creation of the mind of man. The climate of a region is not its "normal" weather in the sense that any departure from this state is unusual. The usefulness of the concept of climate lies in the striking fact that, although the physical properties of the atmosphere undergo large short-period variations, averages taken over periods which are certainly not long in the geophysical sense are remarkably stable. The tempera-

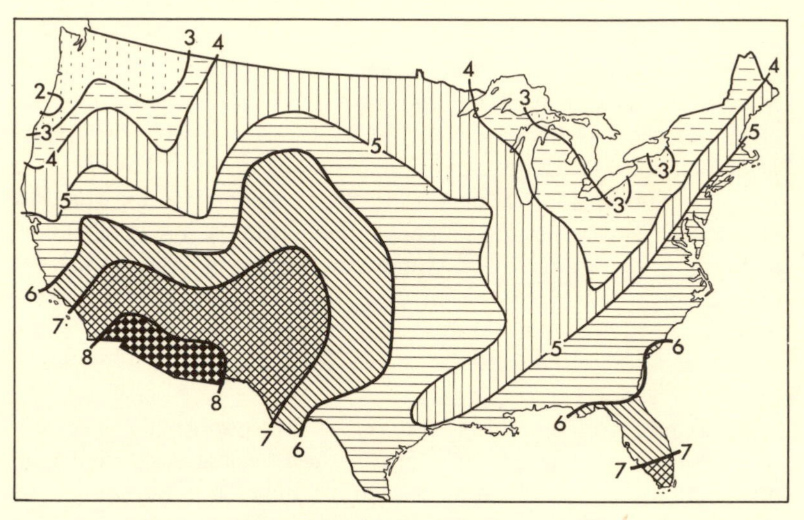

Fig. 1. Normal winter sunshine (number of hours) in the U.S.A. (From *Climatic Atlas of the United States,* ed. S. S. Visher, Harvard University Press. Copyright 1954 by The President and Fellows of Harvard College.)

ture of the air measured near the ground at a meteorological observatory shows relatively little change when averaged over periods as short as a few years, and it is a matter of great difficulty to detect significant trends even in periods as long as a century. Yet day-by-day and month-by-month oscillations of temperature are often large.

Meteorology is often compared, usually to its disadvantage, with astronomy, but the two sciences deal with markedly different systems. Figure 1 shows a typical climatological chart, a pictorial representation of the result of averaging the sunshine records of the United States over a period of years. The chart not only underlines some familiar facts, such as the abundance of winter sunshine in the Southwest and the cloudiness of the Northeast, but it also indicates quantitatively the difference in sunshine amounts between different parts of the country. Charts of this kind are prepared from the records of individual stations, often widely spaced, so

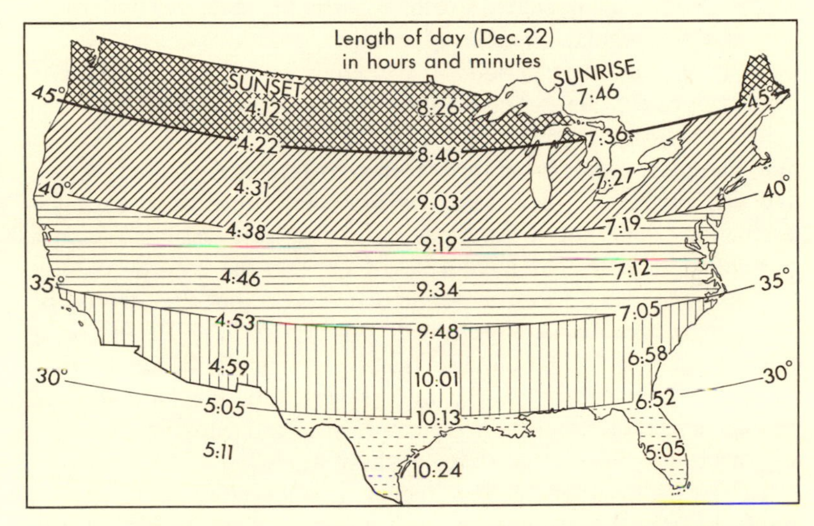

Fig. 2. Length (hours and minutes) of the shortest day (December 22) and the times of sunrise and sunset (local standard time). (From *Climatic Atlas of the United States,* ed. S. S. Visher, Harvard University Press. Copyright 1954 by The President and Fellows of Harvard College.)

that the construction of lines joining points with supposedly equal amounts of sunshine requires judgment, and at times a certain amount of imagination for areas in which observations are scanty. Such lines are in no sense rigidly defined boundaries (like the frontiers of states) between areas of high and low sunshine. They are put in mainly to help the reader to retain the major features of the distribution easily in his memory.

Figure 2 is another chart also dealing with the Sun but this time derived from astronomy. It shows a perfectly regular change in the length of the day with distance from the equator and an equally regular change in the times of sunset from east to west.

The astronomical map is based on calculations involving the geometrical properties of the solar system, and will never change as far as man is concerned. The climatological chart summarizes the experience of a relatively short period, and its preparation did not require calculation of the kind employed in the construction of the astronomical chart. Further, since the data depend on a notoriously irregular element of weather, cloudiness, it is always possible that another period of averaging would show variations from the present chart. This cannot happen in astronomy.

The physical circumstances of the atmosphere, such as the fact that it contains water in all three forms, its relatively shallow depth, and the manner of its heating by the Sun, conspire to produce fluctuations, and not a steady state of the kind assumed in the concept of climate. No meteorologist expects the records of any year to agree exactly with the climatic averages, but the amount of departure varies considerably with position on the Earth. In a country such as Egypt climate is the dominating factor in that variations from the mean are always small. In the British Isles weather constitutes the bulk of experience, and climatic averages are uncertain guides at all times of the year.

To specify the climate of a region it is therefore necessary to know not only the average value of a meteorological element but also its range, or the extent to which it has been observed to depart from the average. Meteorologists, as a rule, are not greatly interested in extreme values, such as "record" temperatures, and prefer to express the range in terms of statistical functions such as the "mean deviation" or the "standard deviation." The reader need not be troubled with the exact definitions of these measures, but it is useful to know, for example, that an observation which does not differ from the long-term average by more than three times the standard deviation usually can be considered as falling within the established climatic range.

These considerations not only are fundamental in any study of climate but also go some way to explain the reluctance of professional meteorologists to associate wet summers or the incidence of hurricanes with man-made disturbances, such as nuclear explosions. Those who have suffered from inclement weather are apt to look around for a scapegoat, and they find little comfort in the assurance that such events are "natural" because they have happened many times before. In advancing such arguments the scientists are doing no more than applying "Occam's razor," the doctrine that "entities are not to be multiplied unnecessarily" enunciated in monkish Latin many centuries ago by William of Occam. In other words, do not bring the fairies unless you have to! All the weather that has been experienced in recent years lies within the known climatic range, and as yet there is no case for the belief that novel influences are at work. It can be truly said that abnormal weather is normal.

Climatic tables and charts are not forecasts but records of things past. Complete regularity and unfailing repetition are found only in astronomy, and a long succession of years with weather which kept closely to the average would be so unusual that there would be substantial grounds for suspecting that some new factor had indeed begun to influence atmospheric processes. There is no single cause of any condition of weather, and even "normal" weather cannot yet be explained fully in terms of measurable properties of the atmosphere.

Although climatological statistics are not always representative of actual states of the atmosphere, obviously there are real differences between the average state of the atmosphere in different parts of the globe. Greenland is justly described as cold and India as hot. There are more subtle and interesting differences between the climate of the United States and that of the British Isles. American weather is in some ways more stereotyped and in other ways more varied than that of Britain. In the Midwest of the United States it is quite certain that the winter will be cold and the summer hot. In England it is sometimes difficult to distinguish between days in December and July except by their length! American weather has a ferocity unknown in Britain, where, as Eliza learned, "*h*urricanes *h*ardly *h*appen." American climate is classified, for the most part, as "continental" with large seasonal variations and characteristic individual phenomena such as tornadoes; the British climate is "maritime," with relatively small seasonal variation, rain and much cloud all the year round but few really destructive storms. What constitutes a "good" climate seems to depend largely on what one has grown up with. The American airman stationed in England is depressed by the frequent rain and fogs and longs for the brilliant sunshine and abundant warmth of his homeland, but the Briton sweltering in Washington in August is apt to remember with nostalgia the gray skies and bracing northeast wind that so often dominate the English summer. Nothing can rival the colors and balmy air of the American fall unless it be the English spring with its flowers—

Daffodils,
That come before the swallow dares, and take
The winds of March with beauty.

The classical treatment of climate gives the greatest weight to three factors: latitude, or position relative to the equator; proximity to seas or large land masses; and topography. These are potent factors in the determination of climate, but today, in assessing the causes of the general weather pattern of a region, the meteorologist is inclined to pay more attention to its position in relation to the major air currents which circle the globe, the so-called *general circulation of the atmosphere*. This is the central theme of dynamic climatology.

This way of studying climate and its problems keeps in view the fundamental truth that climate is the integrated aspect of weather. A scientific study of climate should therefore begin with an inquiry into the physical facts which determine the state of the atmosphere as a whole.

The Heat Balance

The surfaces of the various planets in the solar system have very different average temperatures. The major planets—Jupiter, Saturn, Uranus and Neptune—are very cold, one side of Mercury is very hot, and the surface of Mars is believed to have a mean temperature of about −2° C (28° F). The average temperature of the surface of the Earth is about 15° C (59° F) and this must have had great influence on the evolution of living matter. Can we account for this temperature in terms of the heat received from the Sun, and the physical properties of the atmosphere?

The whole of the energy available to man, whether stored in fuels, locked in the atom, or amassed in the atmosphere, soil, rivers and oceans, is derived, in the last reckoning, from the Sun. The atmosphere is a great heat engine with the Sun as its furnace. Contributions from other heavenly bodies and from the hot interior of the Earth are negligible in comparison with the heat supplied by the solar beam. We know also that the mean temperature of the Earth shows no systematic change when averaged over long periods. This implies that the earth-atmosphere system, in the long run, returns to space exactly as much energy as it receives, and it is the details of the heat balance which present the meteorologist with some of his most formidable and fascinating problems.

The heat of the Sun reaches us as radiation, or electromagnetic waves. The wavelength or frequency of the radiation emitted by a hot body depends upon its temperature, and for the purposes of climatology the Sun may be considered to be a perfect radiator or "black body" (this term has nothing to do with the color of the body) with a surface temperature about 6000° K.[2] It follows from Planck's formula[3] for the radiation from a black body that almost the whole of the Sun's output is in waves of length between 0.17 microns and 4 microns (a micron (μ) is a millionth of a meter). The maximum intensity of sunlight is in radiation of wavelength about 0.5 microns. The human eye responds only to waves between 0.4 and 0.7 microns, so that the peak intensity occurs in the middle of the visible range, the blue-green part of the solar spectrum. Only about half the output of the Sun can be seen by us, but we feel the radiation in the solar beam over a much wider range of wavelength.

Clean dry air is a mechanical mixture of many gases, the most important of which are nitrogen (78%), oxygen (21%), argon (about 1%), and carbon dioxide (0.03%). The proportions are found to be the same for air taken from any part of the globe and

[2] The symbol K indicates the Kelvin or absolute scale of temperature, which for present purposes may be thought of as the celsius or centigrade scale with 273° added. (See Appendix 1.)

[3] See Fig. 4.

they continue unchanged up to very great heights. The atmosphere, however, is never completely dry and rarely clean, and the air in any locality contains very variable amounts of water vapor and pollutants, such as sulfur dioxide. In addition, there are always present large numbers of solid and liquid particles, such as raindrops, ice crystals, salt particles and dust from volcanic eruptions and the erosion of rocks.

The behavior of a gas toward radiation depends upon the nature of the gas and the wavelength of the radiation. Some gases are virtually transparent to radiation of certain wavelengths, that is, the radiation is able to pass through deep layers of the gas without affecting it significantly. The same gas may, however, be opaque to radiation of other wavelengths, in which case the radiation is said to be *absorbed* and the gas is heated. On a clear day the greater part of the solar beam passes through the atmosphere unabsorbed. If this were not so, we should live in a kind of hot gloom by day and intense darkness by night, and the vivid colors of vegetation would not exist. The first step toward the understanding of the working of the atmospheric engine is the realization that the main effect of the Sun's rays is to heat the surface of the Earth, which in turn heats the air in contact with it.

Although sunlight passes through air with relatively little absorption, by no means all the energy provided by the Sun reaches the surface of the Earth. The rate at which energy is supplied to the fringe of the atmosphere, known as the *solar constant,* has been determined by many years of patient observations (in which the Smithsonian Institution has taken the leading part) to be almost exactly 2 calories[4] per square centimeter per minute, or, in the notation used by present-day meteorologists, 2 langleys per minute, a langley being 1 calorie per square centimeter. The solar constant represents an energy income of nearly 140 kilowatts for every square dekameter (an area, say, the size of the average domestic garden) or over 4½ million horsepower per square mile, perpendicular to the solar beam. This is the basic fact on which any attempt to explain the temperature of the Earth must rest.

The solar supply is enormous by any standards but, like all large

[4] The calorie is the scientific unit of the quantity of heat, i.e., of energy. See Appendix I.

incomes, it is subject to severe taxation. At very great heights sunlight acts on the oxygen molecules (O_2) to form ozone (O_3), and it is known that this gas attains its highest concentration between 15 and 20 miles above sea level, where the air is very attenuated. This thin layer of ozone is important both in meteorology and in biology. Ozone is opaque to all solar radiation of wavelength less than 0.3 microns, and no ultraviolet light of wavelength shorter than this is received at sea level. Without the ozone layer the Earth would be exposed to bombardment by short-wave radiation which presumably would have affected the evolution of life.

The amount of energy used in heating the ozone layer is, however, small, and the solar beam suffers its greatest losses in the lowest six or seven miles of the atmosphere, mainly by reflection from clouds and from the surface itself. Clouds are very efficient reflectors of sunbeams, a fact familiar to all who travel by air, and a completely overcast sky means that about three quarters of the incident radiation is reflected into space and completely lost as far as the Earth is concerned. The oceans, which cover about 141 million square miles of the Earth's total surface of about 197 million square miles, also reflect sunlight, as do snow-covered surfaces, glaciers, deserts and vegetation, but clouds are almost certainly responsible for the greater part of the loss.

The fraction of the incoming radiation reflected by a surface is called its *albedo,* and it is clearly a matter of concern to climatologists that the albedo of the whole Earth should be known as accurately as possible. This is not easy, because not only is cloud cover variable but natural surfaces have different albedos. Dry black mold has an albedo of about 0.14 (that is, only about one seventh of the incoming radiation is diffusely reflected) whereas freshly fallen snow rivals cloud sheets in returning nearly three quarters of the energy it receives from the Sun. (It is for this reason that it is possible to have brilliant sunshine and yet firm snow at winter sports resorts.) The simplest and most satisfactory method of finding the albedo would be to measure the reflecting power of the Earth from a point well outside the atmosphere, and with the coming of artificial satellites there is little doubt that this will be done regularly in the future.

In the meantime we must rely on estimates made by climatologists, and in view of the many uncertainties of the problem these naturally cover a considerable range. Values extending from 0.34 to 0.45 have been suggested, all with good reason. For the purposes of a general approach such as the present, a mean value of

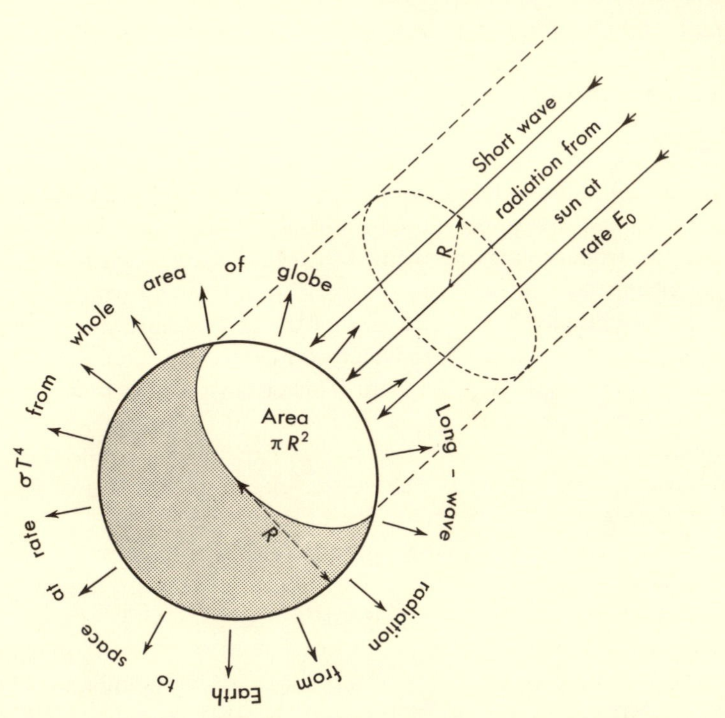

Fig. 3. Calculation of the planetary temperature.

0.4 is adequate, and the arguments which follow would not be changed significantly if another value from the range quoted above were substituted.

With the two constants, the rate at which energy reaches the fringe of the atmosphere (the solar constant) and the albedo of the whole Earth, we can make some simple estimates of terrestrial temperatures. (Fig. 3). For this we need a fundamental law of radiation enunciated by Stefan, that the intensity of radiation emitted by a perfect radiator (a black body) at absolute temperature T is proportional to T^4. In symbols

$$\text{intensity of radiation} = \sigma T^4$$

where σ is Stefan's constant, the value of which is 8.22×10^{-11} langleys per minute.

Meteorological records, and the evidence of history, show that there has been no significant change in the mean temperature of the Earth for many centuries, and it is therefore justifiable to assume that over long periods the earth-atmosphere system returns to space exactly as much energy as it receives. If R is the radius of the Earth and E_0 the solar constant, without the atmosphere an area πR^2 on the surface intercepts radiation at the rate E_0, and if A is the albedo for the whole Earth, the net amount received every minute is $(1 - A)\pi R^2 E_0$. Let us suppose that the earth-atmosphere system emits as a black body at temperature T° K. This temperature (sometimes called the planetary temperature) is that which brings about an exact balance, so that it is determined by the simple equation

$$(1 - A)\pi R^2 E_0 = 4\pi R^2 \sigma T^4$$

since $4\pi R^2$ is the whole surface area of the Earth, which we have assumed to emit radiation as a black body. It follows that

$$T = [(1 - A)E_0 \div 4\sigma]^{1/4}$$

From this simple formula the planetary temperature can be calculated for any value of the albedo A provided that we know the solar constant.

With $A = 0.4$ and $E_0 = 2$ langleys per minute, the calculation gives $T = 246°$ K $= -27°$ C, which is much below the actual surface temperature of our planet. Any increase in the value adopted for the albedo would lower the planetary temperature still more.

The conclusion to be drawn from this simple calculation is that, although we can understand why the surface temperature of the Earth cannot be as low as that of Jupiter or as high as that of Mercury, we still have not found a satisfactory answer to the problem propounded earlier in this chapter. The radiative-equilibrium temperature we have calculated is, in fact, somewhere between the actual mean temperature of the surface and the lowest temperature of the atmosphere which is found, curiously enough, at great heights over the tropics. How, then, are the surface layers of the atmosphere kept warm enough to support life? Clearly, the flow of radiation to space must come partly from the relatively warm surface and the lower layers of the atmosphere and partly from the cold upper air, but there must be a large return flow of radiation from the atmosphere to the surface to keep it warm. Terrestrial radiation must therefore flow both upward and downward in the atmosphere.

There is, in fact, a "greenhouse" effect. The gardener is able to keep his young plants warm by the effective use of glass which, in broad terms, allows sunbeams to enter freely but traps some of the return radiation. The atmosphere acts in a similar fashion, and this brings us to the most difficult but also the most interesting part of the whole problem of the heat balance, the study of radiation exchanges within the atmosphere itself. In particular, we shall see the extraordinary importance of the water-vapor content of the atmosphere in the maintenance of surface temperature.

Terrestrial Radiation and Water Vapor

For the meteorologist, water vapor is the most significant constituent of air. Nearly all the mass of the atmosphere is provided by nitrogen and oxygen, but these gases play only passive parts in the drama of the weather. In meteorology air can be regarded, very largely, as diluted water vapor.

Although water is an exceedingly common substance, it is in many ways peculiar. In the first place, this simple inorganic compound is the only terrestrially abundant substance which is found simultaneously and in the same locality in all three forms—solid, liquid and vapor. Despite the small range of temperature found on the Earth and in its atmosphere, water changes its state frequently. A shallow pool, resulting from the condensation of water vapor into rain, may freeze overnight and dry by evaporation during the next day. Such rapid changes of state are unusual in any natural system, but they are of special importance in the study of weather because of the energy exchanges involved.

A substance can be made to change its state (that is, to pass from the solid to the liquid form, or from the liquid or solid form to the vapor form, or vice versa) only by the supply or release of energy in the form of *latent heat*.[5] Water is peculiar among naturally abundant substances because of its large latent heats. The heat required to melt a gram of ice without change of temperature is 80 calories and to vaporize a gram of water without change of temperature calls for the supply of no less than 600 calories. Both values are large, but the latent heat of vaporization is much higher than that of other common substances. This means that when ice melts, water evaporates, or water vapor condenses, large amounts of energy are absorbed or released. Since these changes of state are the most common physical processes in weather, it is obvious that in the atmosphere energy transfer is on an immense scale, and the water of the atmosphere, by virtue of its unique ability to take up, store and release vast amounts of energy easily, may be regarded as the peculiar ingredient of air which causes weather.

In the present context we are concerned only with the behavior of water vapor toward radiation. As we have seen, the wavelengths of the radiation emitted by a body depend upon its absolute temperature. The surface of the Earth may be regarded, without serious error, as a black body at temperature of about 300° K. According to Planck's formula, this means that terrestrial radiation is composed entirely of waves whose length lies between 3 and 80 microns, with the peak intensity at about 11 microns. Terrestrial radiation is therefore entirely in the infrared or dark part of the electromagnetic spectrum, and there is virtually no overlapping of the solar and terrestrial spectra. There is no risk of confusion if we follow standard meteorological practice and henceforth refer to the incoming radiation as "short" waves and that emitted by the Earth and its atmosphere as "long" waves. The fundamental fact upon which the heat economy of the Earth depends is that, although clear air passes between 80 and 85 per cent of the energy in the short waves from the Sun, it is able, chiefly because of its water-vapor content (and to a lesser degree because of the pres-

[5] See Appendix I.

ence of carbon dioxide), to absorb and re-emit long-wave radiation easily, but the process is not uniform over the whole spectrum.

The study of the absorbing power of a gas as a function of the frequency of incident radiation yields what is known as the *absorption spectrum* of the gas. The absorption spectrum of water vapor in the far infrared is exceedingly complicated, but the essential facts can be grasped without difficulty if we follow the methods of the British meteorologist Sir George Simpson. Although the

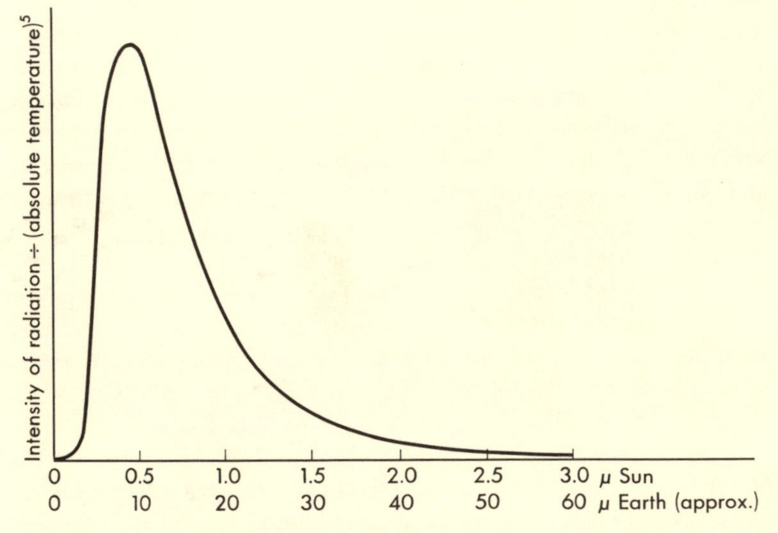

Fig. 4. Intensity of radiation emitted by a black body as a function of wavelength (Planck's formula).

whole of the radiation emitted by a body at temperature 300° K is distributed in waves of length varying between 3 and 80 microns, it is evident from Fig. 4 that by far the greater part of the energy resides in a much less extensive band, say that between 4 and 30 microns. In this reduced range of wavelength Simpson was able to effect a notable simplification of the complex absorption spectrum by some bold assumptions.

The amount of energy absorbed depends, in general, upon the amount of water vapor encountered by the radiation in its passage through the air. Simpson showed that in the lower layers of the atmosphere there is enough water vapor to allow all radiation of wavelengths between 3 and 7 microns, and greater than 14 microns, to be absorbed completely in relatively shallow layers. On the other hand, he treated water vapor as completely transparent to radiation between 8½ and 11 microns. This band of wavelengths is particularly important because terrestrial radiation is at its highest intensity between 10 and 11 microns, and for this reason the gap in the absorption spectrum is often referred to as the "infrared window." The remaining bands, from 7 to 8½ microns and from 11 to 14 microns, Simpson regarded as partially opaque.

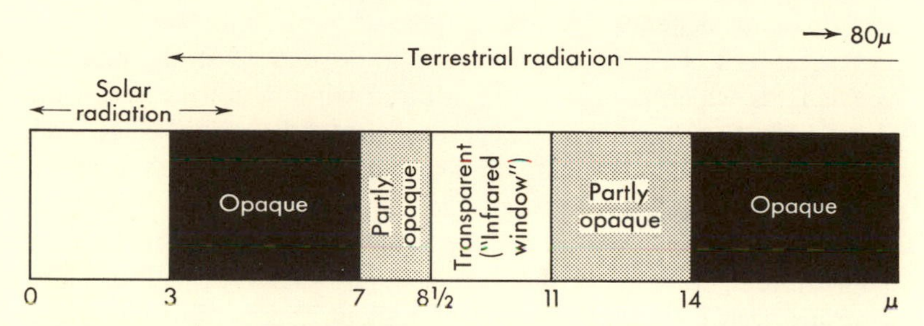

Fig. 5. The absorption spectrum of water vapor in the far infrared as simplified by Simpson.

With this simplification of the complicated absorption spectrum of water vapor (Fig. 5) Simpson was able to demonstrate very simply how the heat balance is achieved by contributions from the various bands. With the albedo of 0.4, the heat income of the Earth is 0.6×2 langleys per minute $\times \pi R^2$, where R, as before, is the radius of the Earth. When this is distributed over the whole surface of the Earth, $4\pi R^2$, the average energy supply is $0.6 \times 2 \times \pi R^2 \div 4\pi R^2 = 0.3$ langleys per minute, and this is the figure at which the heat budget has to be balanced.

To make up the balance sheet we appeal to a fundamental law of radiation first enunciated by Kirchhoff, which, in broad terms, states that a body which is a good absorber (opaque to radiation) is also a good emitter. Suppose that the atmosphere is made up of a series of horizontal layers, each of which contains enough water vapor to allow absorption to be complete in the opaque bands (below 7 microns and above 14 microns). According to Kirchhoff's law, if temperature decreases upward, each layer will absorb energy from the warmer layer below and radiate energy to the colder layer above, and because the intensity of the radiation emitted depends upon the absolute temperature of the layer, the flow of radiation must also decrease upward. Ultimately there will be insufficient water vapor above a layer to absorb radiation, which must then escape to space. The amount lost in this way clearly depends upon the temperature of the final absorbing layer, which is necessarily at a great height above the surface. Simpson's assumption was that the intensity of the radiation escaping in the opaque bands depends upon the temperature of the highest, coldest and driest part of the atmosphere, the stratosphere.[6] In the band of wavelengths which comprise the infrared window all the radiation emitted by the surface will escape to space if the sky is clear, and in the partially opaque bands the amount of radiation lost to space will be about midway between the emission of black bodies at temperatures equal to the highest (surface) and lowest (stratospheric) temperatures in the atmosphere. In a clear atmosphere there are thus two main streams of radiant energy starting at the surface: the stream which passes directly into space through the infrared window and the stream which is continuously being absorbed and re-emitted, ultimately to be lost to space at a rate determined by the temperature of the upper part of the atmosphere.

Simpson devised a simple and attractive graphical method to determine the intensities in the various bands. Here we give only

TABLE 1

Terrestrial Radiation in the Middle Latitudes, after Simpson

Band (microns)	Origin of radiation	Intensity (langleys per minute)
below 7 (opaque)	stratosphere	0.003
7 to 8½	surface + stratosphere	0.024
8½ to 11 (window)	surface	0.079
11 to 14	surface + stratosphere	0.059
above 14 (opaque)	stratosphere	0.128
	Total	0.293

[6] See Appendix I.

the results of his calculations for the mid-latitudes, for which he assumed the mean surface temperature to be 280° K (7° C) and the mean stratospheric temperature to be 218° K (−55° C).

This calculation applies to clear skies. For cloudy skies the temperature of the surface of the earth must be replaced by the average temperature of the upper surface of the cloud sheet, and a mean cloud amount must be assumed. The agreement is satisfactory in this case also.

The water vapor in the air is thus all-important for the maintenance of surface temperature, but before considering the matter further on the global scale it is interesting to see how the greenhouse effect applies at the other end of the scale of size, say, in a California valley. It is a matter of prime importance to the farmer to be warned of the possibility of night frosts during the critical periods in the growth of fruit. There is no simple universal rule that allows night-minimum temperatures to be predicted with certainty, because low temperatures in valleys depend not only upon loss of heat by radiation but also on the possibility of local drifts of cold air to the valley floor from the surrounding hills. However, a tolerable estimate of the night-minimum temperature can be made for a particular locality by taking account of water vapor alone. Figure 6 shows the results of a long series of observations in a fruit-growing district in California. The amount of moisture in the atmosphere may be expressed in several ways. The water vapor of the air exerts a definite partial pressure (known as its vapor pressure) and for a given temperature there is an upper limit to this pressure, called its saturation value.[7] The saturation vapor pressure of water increases rapidly with temperature, a fact which is often loosely but conveniently expressed by the statement that warm air can "hold" more water than cold air. If a volume of moist air is continuously cooled, a point will be reached when the partial pressure of the water vapor it contains reaches its saturation value. This temperature, called the *dew point,* is often used as a measure of the water-vapor content, or humidity, of the air.

[7] It is easy to produce supersaturated vapor with clean air contained in a closed vessel in a laboratory, but in the atmosphere, at temperatures above 0° C, only very small amounts of supersaturation are possible. See Chapter 3.

Another way of expressing the water-vapor content of air is by its *relative humidity,* defined as the ratio of the actual vapor pressure to the saturation vapor pressure at the same temperature, expressed as a percentage.

In Fig. 6 the relative humidity of the air at sunset is plotted against the difference between the dew point at sunset and the lowest temperature reached during the night. The diagram shows

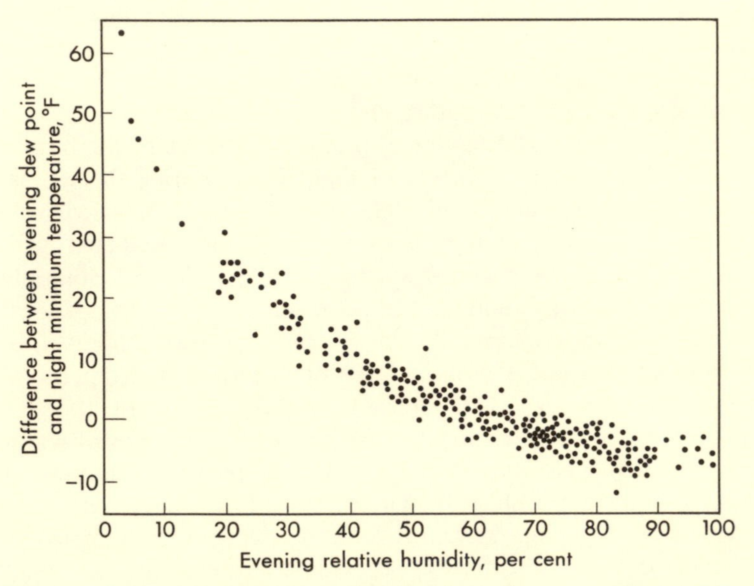

Fig. 6. Minimum temperature forecasting diagram. (From *Micrometeorology* by O. G. Sutton, McGraw-Hill.)

clearly that when the air is dry (low relative humidity) the difference is large, indicating to the forecaster that he must be on the lookout for night frosts. Other factors which contribute to low night-minimum temperatures are clear skies (again because of the loss of heat by radiation to space) and low winds or calms.

To return to the global picture, it is now possible to study the detailed balance sheet for the Earth and the Sun. The estimates quoted here are mainly those of the American meteorologist Henry

G. Houghton, who took 0.34 as the value of the albedo. Let the gross incoming solar energy, which when averaged over the entire surface of the Earth amounts to about 0.5 langleys per minute, be represented by 100 units. With an albedo of 0.34 the net income is then $100 - 34 = 66$ units, which is estimated to be made up of 19 units absorbed by the atmosphere and 47 units by the surface. For the balance, this amount must be shown to be available to be returned to space as long-wave radiation. The surface, emitting as a black body at its mean temperature, produces 120 units, but 106

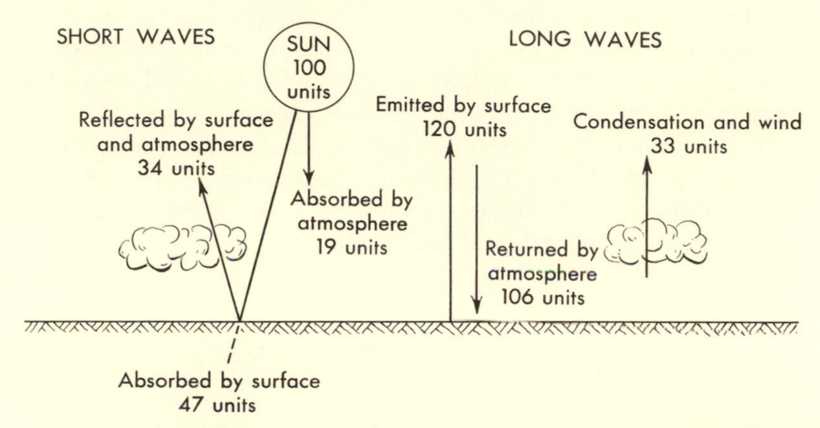

Fig. 7. The heat balance of the earth-atmosphere system and the Sun.

units are estimated to be returned by the water vapor and carbon dioxide of the atmosphere. (This estimate shows the essential part played by water vapor and, to a lesser degree, carbon dioxide in the maintenance of surface temperature.) The difference, called *nocturnal radiation,* is 14 units, and much of this goes into space through the infrared window. Heat is also removed from the surface by the wind; this amounts to 10 units. Condensation of water vapor into cloud releases latent heat, estimated at 23 units, and as we have seen, 19 units are provided by direct absorption of solar radiation. All told, there are thus $14 + 10 + 23 + 19 = 66$ units available, which accounts for the balance. (See Fig. 7.)

The Heat Engine

Meteorology, by an appeal to the well-established laws of physics, coupled with some simplifications and a bold smoothing of the observations, is thus able to account for the heat budget of the Earth and its atmosphere in broad terms. But this is not enough. It is obvious that the equatorial regions receive much more sunshine than the polar regions so that, although the system is balanced as a whole, there must be a surplus in some parts and a deficit in others. This brings us nearer to the root cause ot weather.

The estimation of the heat budget in different parts of the globe is most conveniently done by the division of the hemispheres into 20° zones of latitude, instead of into zones of equal area. The problem is difficult, even in the relatively densely populated Northern Hemisphere, because of the lack of observations, but there is broad agreement between climatologists regarding the main features. Here we shall follow the work of P. Raethjen, whose main conclusions are summarized in Table 2.

TABLE 2

Estimate of the Annual Heat Balance over the Northern Hemisphere, after Raethjen

Latitude zone	Fraction of total area	Short-wave radiation absorbed	Long-wave radiation absorbed	Surplus (+) or deficit (−)
90°–60°	0.14	0.13	0.30	−0.17
60°–40°	0.22	0.23	0.29	0.06
40°–20°	0.30	0.34	0.32	+0.02
20°–0°	0.34	0.39	0.29	+0.10

(units: langleys per minute)

This table brings to light some interesting facts. The short-wave income, as would be expected, decreases considerably from the equator to the pole, but the outward flow of long-wave radiation remains very nearly constant over the whole hemisphere. When the radiation figures are given their proper weight by multiplication by the area fractions in the second column, it is seen that income and expenditure agree at about 0.3 langleys per minute (Raethjen estimated the albedo for the whole Earth to be 0.4) and that surplus and deficit balance within the limits of accuracy imposed by the smoothing. The distribution of radiation given in Table 2 is thus consistent with a perfect balance for the hemisphere as a whole.

The outstanding fact revealed by this and other similar investigations is that there is a surplus of energy income near the equator and a deficit near the poles. If there were no compensating effect, this would result in a steady yearly increase in the temperature of the tropics and a like decrease in the temperature of the polar regions. This does not happen, so that there must be a large-scale transport of heat from low to high latitudes. This is brought about by winds and, to a lesser degree, by ocean currents.

The persistence of climate shows that such compensating movements must follow a fairly regular pattern. We may therefore expect to find in atmospheric motion some enduring features, just as in the ocean there exist well-defined currents, such as the Gulf Stream of the North Atlantic. One such feature, the broad belt of the trade winds, was recognized as far back as the days of sail, and more recently, mainly as a result of improved methods of measuring winds in the upper air coupled with the increasing use of high-flying aircraft, it has been found that there exist, at great heights, at least two persistent narrow belts of very strong winds to which the name *jet streams* has been given.

The broad pattern of air movement over the globe which emerges as the result of averaging the observed winds over long periods of time is called the *general circulation* of the atmosphere. Like all climatic concepts, the general circulation is in one sense a fiction, in that the actual state of motion at any instant may (and usually does) show large departures from the average. With the study of this pattern there has emerged the extremely useful concept of a hierarchy of atmospheric motions distinguished by scale, in which the general circulation is regarded as a basic motion upon which are superimposed motions of successively smaller scale, ranging from those associated with the huge semipermanent continental anticyclones and the familiar cyclonic depressions of the mid-latitudes to thunderstorms, tornadoes, and finally to the most minute of convectional whirls. This aspect of the science of the atmosphere is considered in greater detail in Chapter 2.

The existence of a stable basic pattern of motion underlines the statement made earlier that the atmosphere is a huge heat engine. The essential feature of all such engines is the existence of a "source" and a "sink" of heat. In a steam locomotive the furnace, which vaporizes water in the boiler, is the source of heat, and the cylinder, in which the steam is cooled and made to do useful work by expansion is effectively the sink. In the atmosphere the main source of heat is in the low latitudes and the main sink in the polar regions. The simplest picture we can form of the atmospheric engine is therefore one which resembles the hot-water system of a house, with warm air rising in the tropics, moving to the poles at great heights, sinking and forming a return low-level flow to the equator again. If this picture were strictly true, meteorology would be a delightfully simple science, with surface winds invariably blowing from the north in the Northern Hemisphere and from the south in the Southern Hemisphere, and reverse winds aloft. But even the crudest of observations show that nothing like this happens, and possibly one of the most significant facts in the history of civilization is the existence of a broad belt of very disturbed motion, with predominantly westerly winds, in the mid-latitudes. The existence of this zone cannot be explained on a simple "source and sink" model resembling the domestic hot-water system. The basic reasons for the complexity of the observed motion are well known—chiefly the rotation of the Earth, the irregular distribution of land and water, and most potent of all, the incessant changes in the state of the water content of the air, resulting in the release and absorption of great quantities of energy —but even today there is no completely satisfactory explanation of all the features of the general circulation.

The accumulation of data on the heat balance of the atmosphere and on its general circulation was one of the prime objects of the meteorological program of the International Geophysical Year, 1957–58, and until the great volume of observations has been analyzed it is idle to speculate whether current ideas on the energy budget are sound or will need drastic revision. At this stage we shall therefore leave the problem of the prime cause of weather and consider the hierarchy of dynamical systems in the atmosphere.

CHAPTER 2

THE ATMOSPHERE IN MOTION

As we have seen, meteorologists are now able to present a reasonably convincing account of the energy balance of the atmosphere as a whole. The investigation also brought to light the unequal distribution of the radiation income, as a result of which the atmosphere is in incessant motion over the surface of the globe. The day-by-day changes in the state of the air, which we call weather, are a direct consequence of this activity. The science of meteorology makes great demands upon the techniques of fluid dynamics, but before proceeding to give an account of this work some attention must be paid to the simpler problems of the statics of the atmosphere.

The Troposphere and the Stratosphere

It is commonplace that the temperature of the air usually decreases with height, and it is possible that one of the most surprised men in the history of science was the French meteorologist Teisserenc de Bort when, at the end of the nineteenth century, he found that there was a consistent departure from this seemingly universal rule at heights of the order of 6 or 7 miles over Europe. Until then it had been supposed that air temperature fell continuously with increasing height and would approach absolute zero at the fringes of the atmosphere. At great heights de Bort found a marked discontinuity, and that thereafter temperature ceased to fall and either remained substantially constant or even increased slightly with height.

Figure 8 shows a smoothed representation of the temperature structure of the atmosphere derived from the many observations that have been made in recent years. On the average, temperature decreases at a fairly constant rate up to a height which varies considerably with latitude. In the mid-latitudes a representative

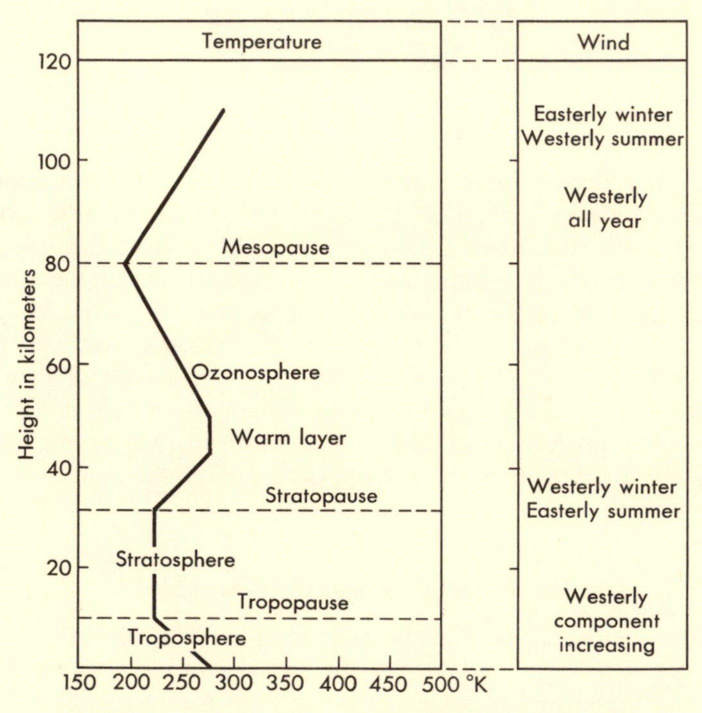

Fig. 8. The temperature structure of the atmosphere.

value for this height is 11 kilometers (7 miles) above sea level. The lower layer of the atmosphere is known as the *troposphere* and the level above which temperature no longer falls is called the *tropopause.*

The troposphere contains about three quarters of the atmosphere by weight and nearly all the moisture and the pollution. The tropopause slopes from the equator to the poles, being about 11 miles high over the tropics and about 5 miles over the poles. These figures, however, are averages; regular soundings show that tropopause height varies considerably and on occasions more than one discontinuity can be identified.

Above the troposphere lies the *stratosphere,* characterized by extreme dryness and uniformity of temperature with height in its lowest layer. The statement of the airline companies that stratospheric flights take place "above the weather" reflects the fact that in the stratosphere there are few clouds and no precipitation. Weather, in the usual sense of the word, is entirely confined to the troposphere.

The atmosphere does not remain at tropopause temperature up to its fringe. Later observations, chiefly on the behavior of meteors and the propagation of sound waves from explosions, have shown that at heights of the order of 30 kilometers (20 miles) the temperature of the attenuated air begins to rise, reaching a maximum value between 50 and 60 kilometers (32 and 37 miles) above sea level. At this height the average temperature of the air lies between 280° K and 290° K (7° and 17° C). After this temperature falls again, to about 190° K (−83° C), at about 80 kilometers (50 miles). At greater heights there is evidence of yet another increase in temperature.

Although it has been claimed that the "the sky is the limit" for meteorology, the atmosphere effectively ceases for most meteorologists at the level of the second temperature minimum, about 80 kilometers (50 miles) above sea level. Some writers consider that the stratosphere should be assigned an upper boundary, the "stratopause," at about 30 kilometers (20 miles) above sea level, with a "thermosphere" or "warm layer" above, but there is no general agreement on the terminology. Above the 80-kilometer level lies the *ionosphere,* in which the molecules of the atmospheric gases are ionized (electrically charged) or dissociated (existing as individual atoms). The word "aeronomy" has been coined to denote the scientific study of this remote region of the atmosphere, but for linguistic reasons this can hardly be regarded as a happy choice.

The causes of the two main increases of temperature in the attenuated high atmosphere are known. Between 35 and 60 kilometers ozone absorbs strongly in the ultraviolet part of the solar spectrum, and above 80 kilometers the D layer of the ionosphere acts in the same way.

The most interesting problem for the meteorologist is, however, to explain the existence of the two distinct lower layers, the troposphere and the stratosphere. The atmosphere, as we have seen, gets most of its heat from below. The density of a gas is directly proportional to its pressure and inversely proportional to its temperature,[1] so that a volume of warm air surrounded by cold air at the same pressure is less dense than its environment and tends to ascend. This motion, which arises because of the buoyancy of the warm air relative to the cold air, is called *natural convection* and is most commonly seen in the rise of smoke from a fire. What is not so obvious is that moist air, which is less dense than dry air at the same temperature and pressure, also may give rise to convection currents.

It must not be thought, however, that convection is bound to occur wherever a fluid is warmer below than above. If the layer of fluid (liquid or gas) concerned is very thin, quite large differences of temperature can be maintained between its upper and lower parts without causing vertical motion, the reason being that in these circumstances the initiation of upward currents is suppressed by the viscosity (internal friction) and the thermal conductivity of the fluid. It has been shown both experimentally and theoretically that before convection starts the difference of temperature between top and bottom must exceed a certain limiting value, depending upon the depth of the layer and the nature of the fluid. This phenomenon, first investigated experimentally by the French scientist Henri Bénard and afterward explained quantitatively by Lord Rayleigh, is of considerable importance in the study of the atmosphere very near the ground, where large differences of temperature in the vertical are a permanent feature of the lowest layers on a hot day.

It is now generally accepted by meteorologists that the explanation of the existence of the troposphere and the stratosphere formulated independently in 1909 by W. J. Humphreys in America and by E. Gold in England is basically correct, although certain points of detail are still not settled. The troposphere is a region in which

[1] See Appendix I.

convection dominates; in the stratosphere heat is exchanged primarily by radiation. The two processes are fundamentally different and lead to dissimilar temperature distributions. Broadly, the observed decrease of temperature with height in the troposphere is caused by the continuous mixing of air from different levels. As air rises it expands because of the reduction of pressure with height. In expanding it cools, and the air that descends to take its place is compressed and heated, but in practice the process is complicated by the evaporation and condensation of water. Mathematical analysis shows that for a dry well-mixed atmosphere to be in equilibrium its temperature must fall at a fixed rate up to a definite height, at which level convection ceases. Above this, according to Humphreys and Gold, the rarefied atmosphere, devoid of convection currents, tends to uniformity of temperature corresponding to a complete balance between radiation and absorption. In this way, by making convection dominant near the surface and radiation higher up, it is possible to account, qualitatively and to a certain extent quantitatively, for the observed features of the real atmosphere. There is little doubt that, basically, the explanation is correct, but the warm ozone layer must have considerable influence on the radiation balance at the tropopause, and there is still much to be explained concerning the radiative balance of the stratosphere as a whole.

In this book it is unnecessary to go into the somewhat complicated mathematics of the theories and it is sufficient for present purposes to regard the troposphere as the part of the atmosphere which is in convective equilibrium and the stratosphere as the part in radiative equilibrium. In meteorological literature the troposphere is often referred to as the region in which the atmosphere is liable to frequent "overturning," but this phrase must not be taken too literally. The atmosphere is not subject to rapid upheavals over large areas, and on the scale of the general circulation the process by which air from the surface layers mixes with air at the higher levels is more aptly described as "infiltration," in the military sense. When averaged over large regions the vertical motion of the atmosphere is extremely small, but because the movement takes place over great areas the volume of air involved is immense and the effects correspondingly significant. Viewed on this scale, the troposphere is like fermenting fluid in a vat; the whole mass is in incessant motion, both up and down and horizontally, and the genesis of the activity is to be found in the ascent of minute bubbles of gas. In the atmosphere vertical currents with speeds comparable with those of the winds are found only in intense local disturbances such as thunderstorms and tornadoes, whose lateral dimensions are very small on the scale of the Earth, or in the vicinity of mountains, when air is forced up a steep slope. The vertical motions which accompany the large weather-producing disturbances of the mid-latitudes (the familiar depressions and anticyclones) are widespread and take the form of the slow rising and settling of huge volumes of air over large areas. The dynamical systems of the atmosphere can be characterized as much by the magnitude of their vertical motions as by their lateral extent. We shall return to this important concept later, and for the present it is sufficient to point out that vertical motion in the troposphere, however small and irregular, is of prime importance. Much of the difficulty experienced in the development of meteorology as an exact (i.e., mathematical) science stems from the fact that, for the most part, vertical motions in the atmosphere are so small compared with horizontal motions that they cannot be measured with sufficient accuracy to allow the values to be used with confidence in numerical work. Yet vertical motion is among the "root causes" of weather and as such cannot be ignored, although in the early stages of the analysis of atmosphere it is frequently necessary to assume that the air always moves horizontally over the Earth in order to obtain a workable mathematical scheme.

Statical Relations

The statics of the atmosphere is concerned, in the main, with the relations between its pressure, temperature and humidity. These properties jointly determine the density of the air, which is rarely measured directly in meteorology, but can be calculated readily from a knowledge of the other three entities.[2] In the troposphere all four properties decrease with height on the average, pressure and density regularly, temperature less regularly, and humidity

[2] See Appendix I.

very irregularly. In the stratosphere, as we have seen, there is little water vapor and temperature at first remains approximately constant, but at high levels shows great variations. Density and pressure, however, decrease regularly to the fringe of the atmosphere.

Meteorologists, unlike the majority of physicists and all chemists, use the absolute unit, the *bar,* instead of the length of a column of mercury, to measure pressure.[2] A barometer is simply a weighing machine, and the pressure of the atmosphere at any level measures the weight of a column of air of unit cross section extending upward from the level of observation to the limit of the atmosphere. It is a simple matter to change from one set of units to another, provided that certain corrections are made to allow for the variation of gravity and the effect of temperature on the length of the column of mercury. In latitude 45°, at 0° C, 1 bar = 1000 millibars (mb) = 750 millimeters = 29.5 inches of mercury. Sea-level pressures are about 1015 mb.

The definition of pressure in terms of the weight of a column of air, when written in the symbols of the calculus, is called the *hydrostatic equation.*[3] This is the basic relation of atmospheric statics. In plain language, the equation states that at any level the rate of change of pressure with height is proportional to the density of the air at that level. Thus pressure must fall more rapidly with height in the dense lower atmosphere than in the attenuated upper air.

There is no simple exact relation between pressure and height above sea level, because the density of the air involves not only pressure but also temperature and, to a lesser extent, humidity, both of which vary irregularly in the atmosphere. However, if the variable temperature is replaced, say, by its mean value throughout the layer concerned, and water vapor is disregarded, it is possible to obtain a straightforward relation between pressure and height which represents the facts with an accuracy adequate for many purposes. Such relations are the basis of barometric altimetry, the method of deducing heights from pressure readings that is used in aircraft. In both theoretical and practical meteorology height above sea level is in many ways an inconvenient parameter, and the meteorologist is now accustomed to regard the atmos-

[3] See Appendix II.

phere as a fluid divided into strata bounded by isobaric (equal pressure) surfaces instead of by level (equal height above sea level) surfaces.

The meteorologist is interested in the ever-changing pattern of atmospheric pressure. The classical method of showing such patterns is to construct *isobars,* or lines joining points with the same pressure, on a level surface. The modern method is to draw *contours,* lines joining points at which a surface of constant pressure has the same height above sea level. By international agreement the pressures used to define the constant-pressure surfaces are standardized at 1000, 850, 700, 500, 400, 300, 200 and 100 mb. The average heights of some of these surfaces are as follows:

Pressure in millibars	Average height in feet of constant-pressure surface
1000	436
700	9880
500	18380
300	30050
200	38360
100	53040

This table underlines the fact that the rate of fall of pressure with height decreases with increasing height. The 100 and 700 mb surfaces are separated by about 9000 feet, but the 500 and 200 mb surfaces are 20,000 feet apart.

The 1000 mb surface is almost at sea level, and the 1000 mb contour chart and the sea-level chart of isobars are very similar in appearance. The 500 mb surface, which is of special importance in theoretical investigations, is situated about halfway between the tropopause and sea level in the mid-latitudes. The 100 mb surface lies in the stratosphere over most of the globe.

The height of a surface of constant pressure changes from point to point and from hour to hour as the pressure changes, and the undulations of the surface show the passage of depressions and anticyclones just as the older isobar charts did. Contours illustrate the absolute topography of a pressure surface, but the meteorologist is equally interested in relative topography, or the vertical distance between two isobaric surfaces, but for another reason. It is easily shown from the hydrostatic equation that the difference in the height of two surfaces of constant pressure, such as the 1000 and

500 mb surfaces, is proportional to the mean temperature of the layer of air separating them. This height difference is referred to as the *thickness* of the layer concerned, and lines joining points with the same thickness are parallel to the *isotherms,* or lines of equal mean temperature. The difference in pressure between two

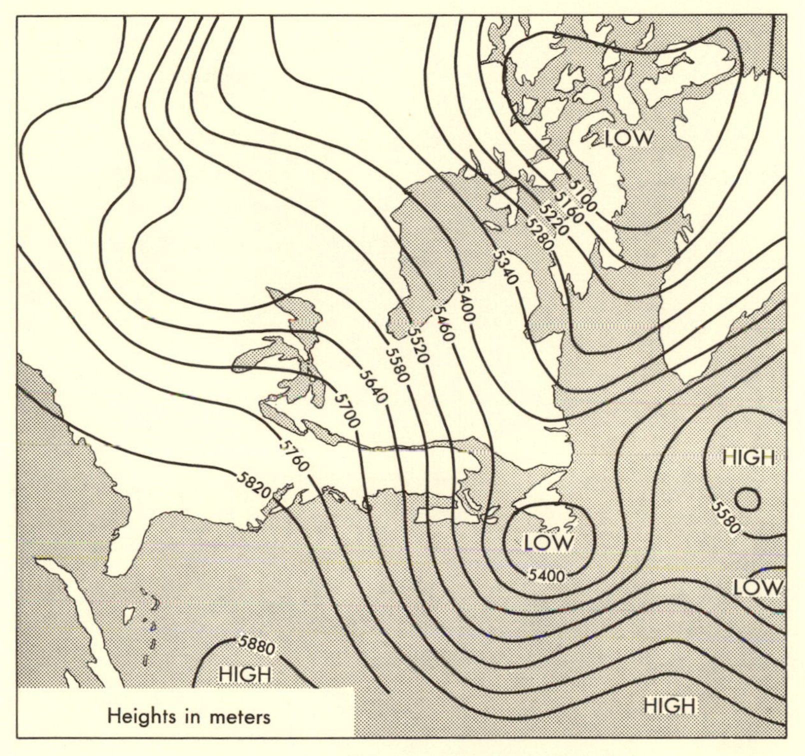

Fig. 9. The contours of a 500 mb surface.

surfaces of constant pressure is a measure of the weight of air between them. Warm air weighs less than cold air, and a large thickness thus indicates a warm layer. Charts of thickness lines have "warm" and "cold" pools just as contour charts have "highs" and "lows" of pressure.

Figures 9 and 10 show two modern aerological charts, one giving the contours of the 500 mb surface and the other the thickness of the 1000–500 mb layer. Such charts form the basis of upper-air forecasting.

Contour and thickness patterns are the framework on which the picture of the dynamical systems of the atmosphere is built, but this cannot be done with pressure data alone. It is necessary

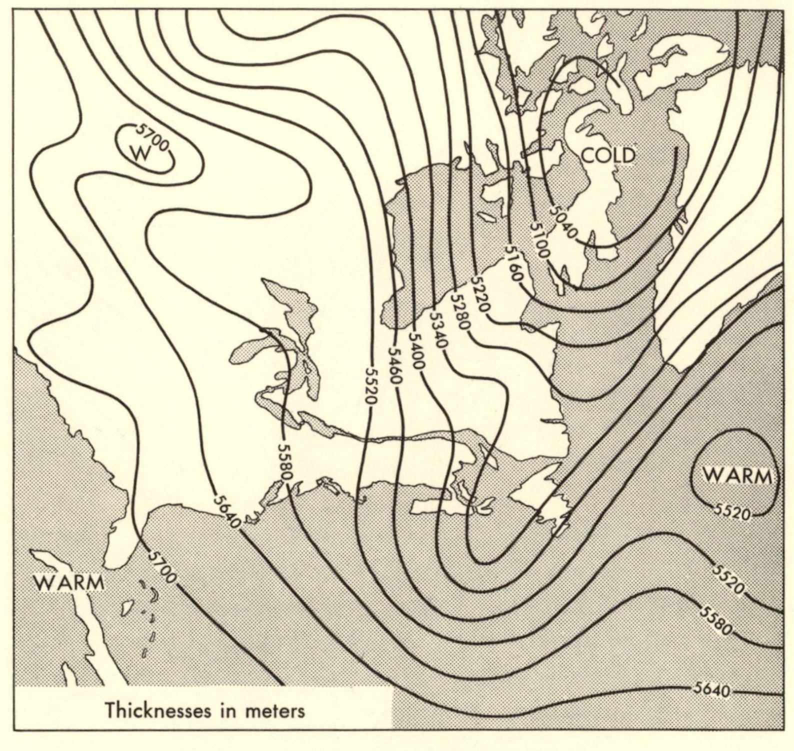

Fig. 10. A 1000–500 mb thickness pattern.

to take into account changes of temperature, especially those which occur between the surface and the tropopause.

Lapse Rates and Stability

The troposphere is the region of the atmosphere in which, on the average, temperature falls with height at a rate which keeps fairly close to 0.6° C in 100 meters (0.3° F in 100 feet). Is it possible to deduce this fact from the principles of physics without recourse to measurements made in the atmosphere itself? The answer is that we can approach fairly closely to the observed value by purely deductive arguments, provided that we make a fundamental assumption concerning the manner in which heat is transferred in the majority of atmospheric processes.

The assumption in question is that the processes involved in the large-scale dynamical systems are *adiabatic,*[4] that is, the temperature of a volume of air in motion rises or falls without transfer of heat across its boundaries. If radiation, condensation and evaporation are excluded, this means that changes in temperature are caused solely by changes in pressure. The most common example of an adiabatic process is the heating of the cylinder of a pump when an automobile tire is being inflated, and there is a similar cooling effect when the pressure of a volume of air is suddenly lowered. The essential feature in both examples is that the compression or expansion takes place so rapidly that only a negligible amount of heat is conducted across the boundaries of the volume.

If a volume of air is forced, for some reason or other, to rise or fall in the atmosphere, it must undergo a change of pressure. Adjustments for pressure are almost instantaneous, but the conduction of heat is a much slower process. It follows that any analysis of vertical movements in the atmosphere should not depart too far from reality if the assumption is made that all changes of temperature resulting from changes in level are adiabatic. This allows the mathematical treatment to be greatly simplified.

The troposphere is always being stirred by the winds and there is a continuous, but slow, exchange of air between one level and another. If a small volume of fluid is well stirred, the effect is to produce uniformity of temperature—an *isothermal* state. The usual way of keeping a body at a steady temperature in a laboratory is to immerse it in a water bath agitated by a paddle. Stirring of the atmosphere, however, produces not an isothermal state but one in which temperature decreases at a fixed rate with height above sea level. The reason is that any change of level in the atmosphere is accompanied by a change of pressure and an adiabatic change of

[4] See Appendix I.

temperature, and because of the great depth of the atmosphere these changes are significant. In the laboratory water bath the differences in pressure are small and their total effect negligible. In meteorology the rate of decrease of temperature with height is called the *lapse rate,* and the basic problem is thus to calculate the lapse rate of a well-stirred atmosphere. The problem may be recast in a more illuminating form if it is related to the *stability* of a layer of air in which vertical currents exist.

A system is said to be in a stable equilibrium if, after a small change, it reverts of its own accord to the original state. In a stable atmosphere a small volume of air displaced upward or downward experiences forces which act to restore it to its original level; in an unstable atmosphere air displaced vertically tends to continue rising or falling, motions which if maintained indefinitely would cause complete overturning. The stabilizing or destabilizing force involved is buoyancy, which depends on the density of the air relative to its environment. We are thus led to relate stability to the temperature changes of a rising or falling volume of air.

The calculation of the rate of change of temperature of a volume of air which expands as a result of upward motion, or is compressed by being forced downward, is not difficult if the process is supposed to be adiabatic and condensation and evaporation are excluded. This rate of change is found to be g/Jc_p, where g is the acceleration due to gravity, J is the mechanical equivalent of heat, and c_p is the specific heat of dry air at constant pressure. The numerical value of this quantity, known in meteorology as the *dry adiabatic lapse rate,* is approximately 1° C per 100 meters or about 0.5° F per 100 feet. The dry adiabatic lapse rate, denoted by the symbol Γ, is one of the fundamental constants of meteorology.

This calculation shows how it is possible to decide from observations of the rate of change of temperature with height whether or not an atmosphere is in stable equilibrium. If all processes are supposed to be adiabatic and there is no release or absorption of latent heat by condensation or evaporation, the temperature changes of the rising or falling air are entirely caused by changes of pressure and are thus fixed by the adiabatic lapse rate. The stability or instability of the atmosphere then depends upon the difference between the actual and the adiabatic lapse rates.

If the observed lapse rate in a dry atmosphere is equal to the adiabatic value, a volume of air moving up or down always is bound to have the same temperature and pressure, and therefore the same density, as its surroundings. The stability is then said to be *neutral*. If the observed lapse rate exceeds the dry-adiabatic value, a volume of air displaced upward expands and cools but is always warmer, and thus less dense, than the surrounding air, and if moved downward is compressed and heated but is always cooler, and thus denser, than its environment. In both instances the buoyancy forces created favor the continuance of the vertical motion, and the layer of air concerned is said to be *unstable*. Thunderstorms, tornadoes, and other manifestations of strong convection are associated with superadiabatic lapse rates. On the other hand, the observed lapse rate may be less than the dry-adiabatic rate or even reversed in sign (temperature increasing with height in the layer concerned, a condition frequently met in the atmosphere and known in meteorology as an *inversion*). A volume of air displaced upward is then always cooler (denser) than the surrounding air and if moved down is warmer (less dense) than its environment. In both instances the displaced air is subject to a buoyancy force which tries to return it to its original level, convection is suppressed, and the layer is said to be *stable*. Such conditions are associated with anticyclones and, only too often, with fog. The magnitude and sign of the rate of change of temperature with height—the vertical temperature gradient, as it is called—is thus important in forecasting, and for this reason the temperature-height curves are plotted from the observations on special thermodynamic charts on which lines corresponding to the dry-adiabatic lapse rate are prominently displayed as guides to the forecaster.

The deduced value of 1° C per 100 meters is close to the observed average value of the lapse rate for the troposphere, 0.6° C per 100 meters, a fact which supports the hypothesis that the majority of changes involved are adiabatic. Complete agreement between calculation and measurement is not to be expected, because the real atmosphere is not dry and some nonadiabatic changes must take place. When condensation occurs in moist air as a result of adiabatic expansion, latent heat is released and the rate of cooling of the rising air is reduced. The rate of fall of temperature of saturated air subject to adiabatic cooling can be calculated without difficulty, but the saturated adiabatic lapse rate, as it is called, is not constant but depends upon the temperature of the air. In warm tropical air the dry and wet rates differ considerably, but the difference is negligible in the cold upper air and in the polar regions. The results of the investigation thus suggest the important conclusion that the troposphere is never far removed from instability. On the other hand, the cold dry lower stratosphere is clearly indicated as a region of stability.

The appearance of a superadiabatic lapse rate in a layer of the atmosphere is not a sure sign that severe weather is imminent, but it puts the forecaster on the alert for thunderstorms, and perhaps tornadoes. Much depends upon the depth of the layer, its moisture content, and where it is found. Superadiabatic lapse rates are a permanent feature of the air very near the ground in daytime, except when the sky is overcast. The largest values occur within a few inches of the ground in clear summer weather. A difference of nearly 10° C has been recorded between screened thermometers placed at 1 inch and 1 foot above a lawn during a sunny afternoon in England. This difference, if interpreted as a lapse rate of 10° C per foot, is equivalent to nearly 2000 times the dry-adiabatic lapse rate, but this state is peculiar to thin layers very close to the ground. Lapse rates of such orders of magnitude are never found in deep layers well above the surface.

The Laws of Atmospheric Motion

Although fluid dynamics is a notoriously difficult branch of natural science, it is founded upon relatively simple principles. One, which is almost self-evident, is that a difference in pressure between two points in the fluid in the same horizontal plane sets up a force which drives the fluid from the higher to the lower pressure. This happens when water flows along a pipe, and in the atmosphere wind is caused by differences in pressure over the surface of the globe. More precisely, the force so created is proportional to the *pressure gradient,*[5] or the difference in pressure divided by the distance separating the points, and the direction of flow is usually "down the gradient."

A cursory inspection of weather maps for nonequatorial regions suggests that the rule does not hold in the atmosphere, for, as can be seen from Fig. 11, winds blow around centers of low pressure,

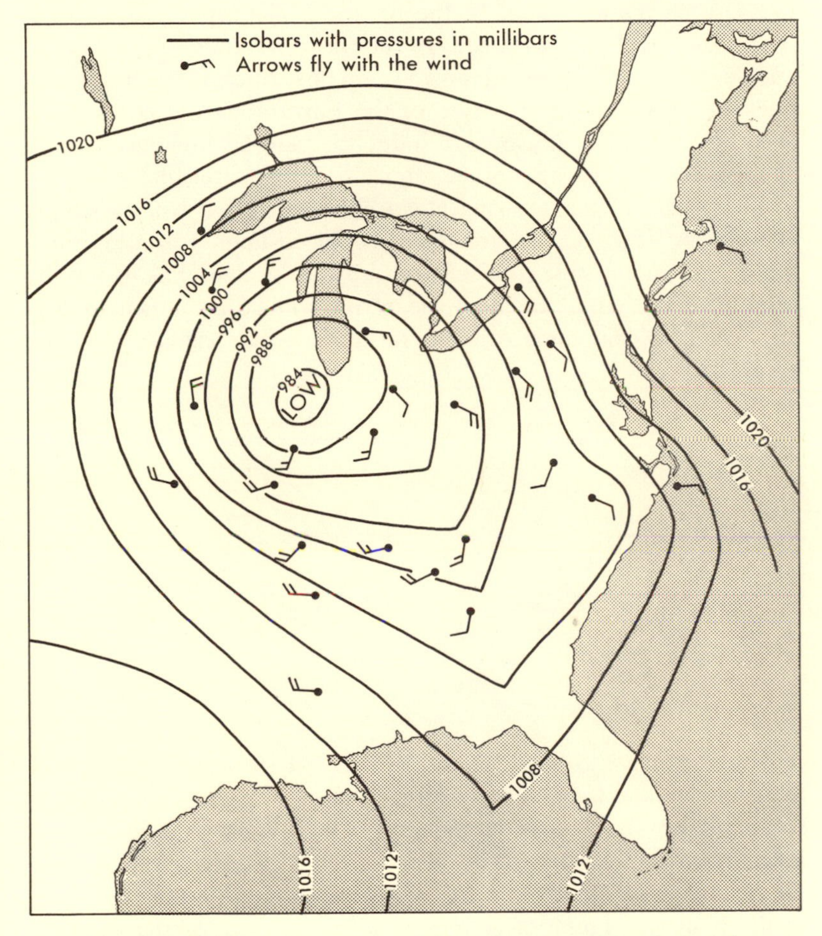

Fig. 11. Winds and pressure fields (Northern Hemisphere).

[5] See Appendix I.

keeping more or less *along* the isobars, whereas the direction of the pressure gradient is *across* the isobars. It is evident that a steering effect has intervened to force the air particles to move nearly at right angles to the pressure gradient. In the Northern Hemisphere the direction of the wind is found from experience to conform to the famous rule known as Buys Ballot's law: that an observer standing with his back to the wind has the lower pressure on his left hand. In the Southern Hemisphere the rule is reversed—the lower pressure is on the observer's right hand—and it does not hold at all on the equator. Such a steering effect must be caused by the rotation of the Earth about its north-south axis, for no other dynamical influence can be imagined which changes sign on crossing the equator.

The nature of this steering effect can be realized from a simple experiment. If a sheet of stiff paper is placed upon a rotating turntable and an attempt is made to draw a straight line from the spindle to the circumference, the result will be as shown in Fig. 12. An intelligent insect on the turntable would deduce that the

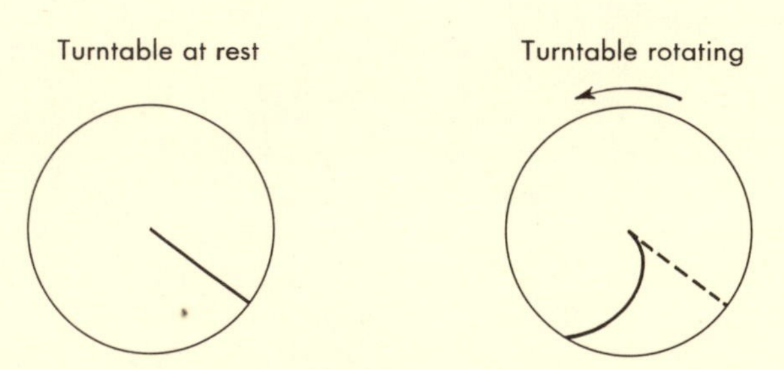

Fig. 12. The effect of the rotation of the Earth on winds.

crayon drawing the line was subject to two forces, one directed from the spindle to the circumference along a radius and another force, acting at an angle to the first, which causes the crayon to deviate from the radius. The observer, who is not rotating with the turntable, would explain the result by saying that the curvature of the path was simply a geometrical consequence of the rotation of the turntable. Both would be correct in the sense that either view could lead to correct predictions of the path.

To apply this to the atmosphere, the spindle may be taken to represent a pole and the circumference of the turntable the equator. A balloon drifting at a constant speed horizontally from the North Pole along a line whose direction is fixed in space thus appears to an observer on the ground to be deflected to the right, or to be coming from east of north. To account for this effect the observer might postulate the existence of a force acting from left to right on the balloon. In meteorology, winds are always described relative to the Earth and it is convenient to introduce such a force to account for the effect of the rotation of the Earth about its axis.

The fictitious force introduced in this way is known as the *Coriolis,* or *deviating force.* The mathematical analysis of the problem shows that the expression for the force D is

$$D = 2V \rho \omega \sin \phi$$

where V is the speed at which the particle moves relative to the Earth (in our case, the wind speed), ρ is the density of the air, ω is the constant angular speed of rotation of the Earth about its axis, and ϕ is latitude. Clearly, the force is a maximum at the poles ($\phi = 90°$) and zero on the equator ($\phi = 0°$).

The Coriolis force acts on all bodies moving freely over the surface of the Earth. It can be detected in long-range artillery bombardments but is much too small to have an appreciable effect on the flight of a golf ball. In meteorology, the deviating force exercises a first-order influence in large-scale motions, such as those which produce weather, but it has little effect on local surface drifts of air which arise mainly from topographical effects.

When a horizontal pressure difference exists in the atmosphere the air at first moves "down the gradient," but as it acquires velocity it comes increasingly under the influence of the Coriolis force and deviates to the right (in the Northern Hemisphere). Within a relatively short time the pressure-gradient force and the deviating force come into balance and the motion of the air becomes *steady,* that is, it does not change with time when measured at a fixed point. It can be shown that in an idealized system, in which the pressure field is stationary and there is no friction, there is a unique velocity of the air which achieves the balance. This velocity, known as the *geostrophic wind* (Fig. 13), is proportional to the magnitude of the pressure gradient and is directed along the isobars. If the isobars have no appreciable curvature, the magnitude of the geostrophic wind V_g, is given by the simple formula:

$$V_g = \text{pressure gradient} \div 2\,\rho\,\omega \sin \phi$$

If the isobars are not straight but curved, as happens near the center of a depression, a correction must be introduced to allow for the acceleration toward the center caused by the air being constrained to move on a circular path. The motion which results

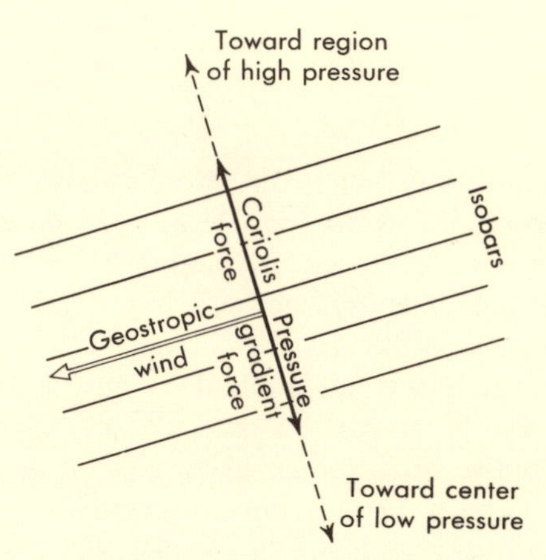

Fig. 13. The geostrophic wind in the Northern Hemisphere.

is called the "gradient wind," but for present purposes we may disregard this complication and consider only geostrophic winds.

One of the most remarkable and significant facts of meteorology is that in layers of the atmosphere outside the tropics in which the friction of the surface is not appreciable the motion of the air in large-scale systems conforms closely, but not exactly, to the geostrophic balance. This result is indispensable in operational meteorology. As a working approximation, the forecaster considers the wind at heights of the order of 2,000 feet to blow along the isobars of the surface pressure field with the geostrophic speed. The pressure gradient is inversely proportional to the spacing of the isobars (or contour lines) of the synoptic chart, and it is a simple matter to construct a transparent scale which when placed across the isobars allows the geostrophic speed to be read off without calculation. For the operational meteorologist the geostrophic wind is the link between the pressure and motion fields of the atmosphere but, like all approximations, it must be used with knowledge. In equatorial regions, where sin ϕ becomes very small, the geostrophic balance has no meaning, and in the low latitudes there is often no clear-cut relation between the surface-pressure distribution and the motion of the air. Near the ground, friction causes the wind to blow at an angle to the isobars, inclined toward low pressure at a speed appreciably less than the geostrophic value. The amount of deviation depends upon the roughness of the surface, and on the average, surface winds blow at about 25° to the direction of the isobars at speeds between one half and two thirds of the geostrophic value.

Thermal Winds

One of the earliest discoveries in meteorology was that wind speed and direction usually vary considerably with height. It frequently happens that the wind field at the heights at which modern aircraft operate bears little resemblance to that in the lower layers. The relation between the two motion fields is conveniently expressed by the concept of the *thermal wind.*

Suppose that the geostrophic balance is satisfied at all levels in the atmosphere. To fix ideas, consider the 1000 and 500 mb surfaces which, as we have seen, are situated approximately at sea level and 18,000 feet, respectively. Figure 14 shows an idealized pattern of the heights (contours) of the two surfaces. Contour lines pass through points having the same pressure, so that contours are also isobars, and at each level the geostrophic wind blows along the contours at a speed inversely proportional to the spacing of the lines. The difference in height between the two isobaric surfaces is indicated by the broken lines drawn through the vertices of the quadrilaterals formed by the two sets of contours; these are thickness lines which, as we have seen (p. 31), are also isotherms of the mean temperature of the layer of air between the constant pressure surfaces. It is easily seen that the difference between the geostrophic winds on the isobaric surfaces is a motion along the thickness lines of magnitude inversely proportional to the spacing of the lines (i.e., directly proportional to the horizontal gradient of mean temperature in the layer). This motion is the thermal wind V_T. Its magnitude is given by

$$V_T = g \times \text{thickness gradient} \div 2\,\omega \sin \phi$$

where g is the acceleration due to gravity.

In the Northern Hemisphere the thermal wind blows along the thickness lines so that the lower temperature is on its left, just as

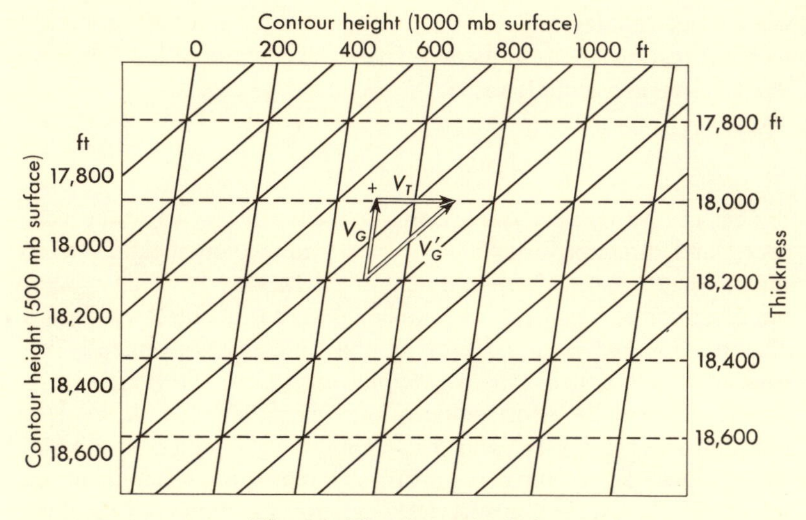

Fig. 14. The thermal wind.

the geostrophic wind blows along the contour lines, or the isobars, keeping the lower pressure on its left. Buys Ballot's law thus applies to thermal winds if we simply substitute temperature for pressure.

The concept of the thermal wind as the link between the upper and lower pressure fields has proved extremely useful both in the practical problem of forecasting and in theoretical studies. Today meteorologists think of the geostrophic wind in the upper air as the sum of the surface geostrophic wind and the thermal wind, but in combining the two motions attention must be paid to both speed and direction—the addition has to be done "vectorially." In precise terms, the thermal wind represents the difference between the geostrophic winds at two levels; or, because in a geostrophically balanced system isobars are also streamlines,[6] the upper-air pressure field shows the influence of the temperature of the intervening layers on the surface-pressure field. As will be seen later, these concepts have had considerable influence on investigations into the behavior of large-scale dynamical systems in the atmosphere.

The picture we have drawn of atmospheric motion as the result of a complete balance between the pressure gradient and the deviating forces has one important (and rather disturbing) consequence. Vertical motion has no place in a system in geostrophical balance. This is often summed up in the phrase that with rigidly geostrophic winds there could be no "weather," for clouds and precipitation depend essentially on vertical motion. Departures from the geostrophic balance are therefore of the utmost importance for the meteorologist. The fact the geostrophic winds approximate closely to actual winds is useful in operational meteorology, but can be an embarrassment in mathematical work. It carries with it the implication that the cross-isobaric flow, which is all-important in the problem of the growth or decay of the large-scale motion systems, is small and therefore difficult to evaluate. A strict geostrophic motion system is sterile in the mathematical sense, but fortunately it has been found possible to make use of the balance principle without restricting the applicability of the results too severely. We shall deal with this point at greater length in a later chapter.

Dynamical Systems of the Atmosphere

An instantaneous picture of the movements of the whole atmosphere, if it could be made, would be so complicated as to be

[6] A streamline has the same direction as the velocity of the fluid at every point. If the motion is steady, a streamline is also the path of a particle of the fluid.

meaningless. To make order out of chaos it is necessary to average the winds over several periods of time and to divide the atmosphere into several levels.

The formation of averages over different periods of time is a kind of filtering process. The record of a sensitive anemometer placed a few feet above the ground shows that the natural wind is made up of a rapid succession of gusts and lulls. This type of motion is called *turbulent,* and may be thought of as the passage of whirls and eddies, ranging from fractions of an inch to many feet in diameter, past the anemometer head. If the record is averaged over successive periods of an hour, much of the irregularity of the original motion disappears, leaving a relatively smooth mean flow. The total motion may thus be thought of as a mean flow on which are superimposed secondary motions caused by the eddies. The mean flow is typical of a small area around the anemometer and shows only the passage of medium- and large-size systems, such as thunderstorms and depressions. The small-scale fluctuations are filtered out by the process of time-averaging.

The synoptic charts[7] used for routine weather forecasts contain observations of the fluctuating meteorological elements averaged over a period of about an hour. The analyst, by drawing isobars and isotherms, carries out a further process of spatial averaging. When, in the nineteenth century, the invention of the electric telegraph made synoptic weather charts possible, it soon became evident that there exist in the atmosphere clearly defined individual circulations, or dynamical systems, which move as separate entities and have "diameters," or horizontal scales, varying from hundreds to thousands of miles. On a modern synoptic chart, constructed from observations collected from the meteorological network in settled areas such as North America or Europe, it is possible for a skilled analyst to identify individual circulations with horizontal scales down to a few tens of miles. Although smaller systems are known to exist, they are usually below the "resolving power" of the normal meteorological network, and in parts of the world where observations are few only the major systems can be identified with certainty.

[7] See Appendix I.

Most of the circulations shown on the official nationwide weather charts are the familiar lows (depressions) and highs (anticyclones). When the period of averaging is extended to days or weeks and the area of observation enlarged to embrace the whole of, say, the Northern Hemisphere, individual systems such as lows are filtered out in turn, leaving a smooth flow with well-defined characteristics. This basic motion is the general circulation of the atmosphere.

The driving force of the general circulation is provided by the Sun. The solar beam increases the temperature and therefore the internal energy[8] of the atmosphere, which is proportional to its gravitational potential energy. Some of this potential energy is transformed into kinetic energy, or energy of motion. In general, this transformation takes place by expansion and contraction, which in the atmosphere are linked with vertical motion. Because of the gravitational field, rising air must be less dense, on the average, than descending air, which means that the center of gravity of the whole system is lowered. The unequal heating of the surface of the Earth is the essential preliminary to these processes, since regions of ascending air obviously must be adjacent to regions of descending air. The characteristic phenomena of weather arise because vertical motion is accompanied by condensation and evaporation, which, however, greatly complicate the transformation processes by the release or absorption of large amounts of energy.

In an examination of atmospheric movements the scale aspect looms large. The basic scale of length must be the "depth" of the atmosphere, which, for present purposes, we may take to be of the order of tens of kilometers (or miles). (Only the power of 10 is significant.) With this criterion in mind we may classify the motion systems of the atmosphere as shown on p. 46.

The dynamical systems listed there are distinguishable not only by their horizontal scales but also by the nature of the air movements within them, especially in the vertical. On the scale of the general circulation the mean vertical motion is very small, prob-

[8] The internal energy of a gas is measured by the product of its absolute temperature and its specific heat at constant volume.

Horizontal scale	System	Operational significance
Very large (thousands of miles)	General circulation (trade winds, continental anti-cyclones, etc.) Long waves	Long-range forecasts (month or more ahead)
Large (thousands to hundreds of miles)	Weather-producing systems (depressions, small anti-cyclones) Short (unstable) waves	Short- and extended-range forecasts (day or two ahead)
Medium (tens of miles)	Meso-scale disturbances (thunderstorms, tornadoes)	Very short-range fore-casts and warnings
Very small (miles or yards)	Eddies (diffusion, local winds, frosts)	Advisory services for agriculture, etc.

ably not more than a hundredth of a foot a second. The basic motion of the atmosphere is thus very nearly horizontal. This is true also of winds in the large depressions and anticyclones, in which the average rates of ascent and descent of air are a little larger, perhaps a few tenths of a foot a second. It is only in relatively small intense circulations such as hurricanes, tornadoes and thunderstorms that substantial vertical currents are found. In a celebrated investigation into the structure of thunderstorms carried out in Texas, upward currents up to a hundred feet a second were measured in the interior of the cumulus clouds, but such motions are short-lived and exist over only relatively small areas. The vertical component of velocity is of the same order of magnitude as the horizontal components in the tiny frictional eddies in shallow layers near the surface of the Earth, but here the speeds are always lower than those which characterize deep convection phenomena such as thunderstorms and tornadoes. In general, the larger the dynamical system the weaker its vertical motion, a feature of atmospheric flow which is of the greatest importance in the development of meteorology as an exact science.

The General Circulation

We conclude this chapter with a brief account of the problem of the general circulation of the atmosphere. The aim of the meteorologist here is to explain quantitatively the origin and maintenance of the major wind systems of the globe. So far this has not been fully achieved, and the mathematical difficulties standing in the way of a complete solution make it impossible, in a book of this kind, to attempt more than a broad outline of the observed features and the theories which have been evolved to account for them.

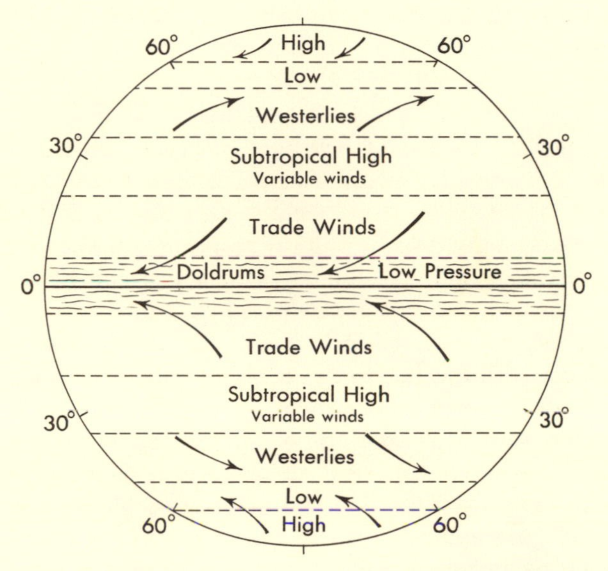

Fig. 15. The basic pattern of surface wind and pressure fields over the Earth.

It is convenient in this problem to divide the total motion of the air into three components: parallel to lines of latitude (zonal) and to lines of longitude (meridional) and vertical. The mean motion is nearly geostrophic, so that it is immaterial whether we regard the pressure or the motion field as basic.

Figure 15 is the simplest possible pattern of the surface pressures and winds, such as would be expected to occur on a uniform Earth. Near the equator is a belt of low, uniform pressure, characterized by calms and light variable winds, occasionally broken by squalls, heavy downpours of rain, and thunderstorms. This region, originally known as the *doldrums,* is now usually referred to as the *intertropical convergence zone*. It marks the boundary between two zones in which the winds are unusually steady, from the northeast in the Northern Hemisphere and from the southeast in the Southern Hemisphere. These are the trade winds, which played such an important part in the opening up of the New World by mariners from Europe. At about latitude 30° lie two belts of high pressure with generally light winds—these are the *subtropical high-pressure zones*. Between latitudes 35° to 60° are the zones of the *westerlies,* marked by abundant depressions, changeable weather, and generally temperate conditions. Farther north and south are other well-defined belts, the *subpolar lows,* and finally the *polar high-pressure areas*. The zonal motion (along lines of latitude) is the dominant component, with a marked increase with height in the westerly direction in the troposphere. The meridional component (along lines of longitude) is much smaller, and the vertical component very small compared with the zonal motion.

Because of gravitational attraction, the atmosphere, on the whole, rotates with the Earth, but it also exhibits internal motions (winds) resulting from the irregular heating of the surface of the globe by the Sun. The problem of the general circulation is to formulate a consistent physical theory which, starting with the known heat input from the Sun and taking into account the astronomical and geographical features of our planet, can explain the genesis and maintenance of the circulation. At present this is beyond the powers of science, for the problem as stated is too complicated for mathematical analysis and not all the facts are known. It is possible, however, to give a broad explanation of how the atmospheric engine works, much in the way in which the working of an automobile engine could be explained to a nontechnical user. We cannot yet explain all the details, any more than the average automobile owner can explain, for example, the intricacies of the latest automatic gear-change mechanism.

The basic reason why the air moves relative to the Earth, the unequal heating of the globe by the Sun, was first realized by Edmund Halley (of comet fame) and published by him in 1686 in a celebrated essay on the trade winds. A better account was given in 1735 by another Englishman, George Hadley, who suggested that air ascends near the equator and descends in higher latitudes to give a meridional (poleward) drift in the upper air, with a return equatorial current at the surface, which is deflected by the rotation of the Earth into the northeast and southeast trades. Such a circulation (Fig. 16) is now called a *Hadley cell,* and there is abundant evidence that motion of this kind occurs between the equatorial regions and latitudes 30–35°.

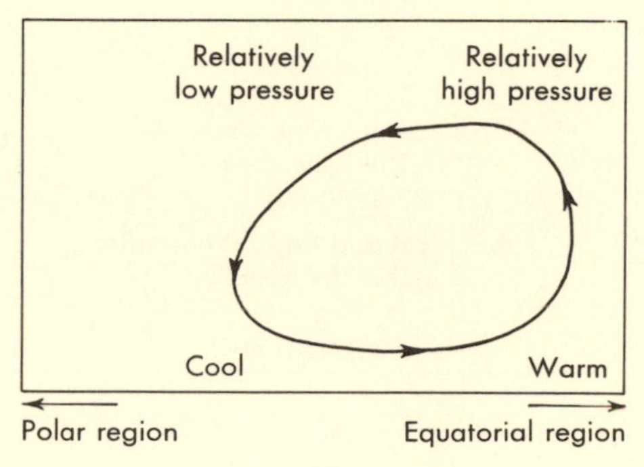

Fig. 16. Hadley cell.

This explains the belt of surface easterlies in the tropics, and it would be highly satisfactory if the surface winds were easterly everywhere. But this is impossible, for the rough surface of the Earth must exert a large frictional drag on the air, and any zonal surface current which is predominantly in one direction must be compensated by an opposing zonal motion elsewhere if the atmospheric envelope and the Earth are to continue to rotate at a rate which, for the purposes of meteorology, never changes. To maintain this steady state, belts of generally easterly winds must be accompanied by westerly motion elsewhere on the globe.

To explain the mechanics of the system further necessitates bringing in another concept of dynamics, *angular momentum.* Newton introduced momentum to measure the "quantity" of motion. For a particle of mass m, moving in a straight line with speed v, momentum is simply mv. Angular momentum is the same concept applied to a particle rotating about an axis, and is proportional to the angular velocity of the particle and the square of its distance from the axis of rotation. In the tropics, where distance from the axis of rotation of the Earth is large, air particles have much greater angular momentum than particles near the poles, where distance from the axis of rotation is less.

Because of surface friction (which always acts against the motion), easterly winds acquire westerly angular momentum from

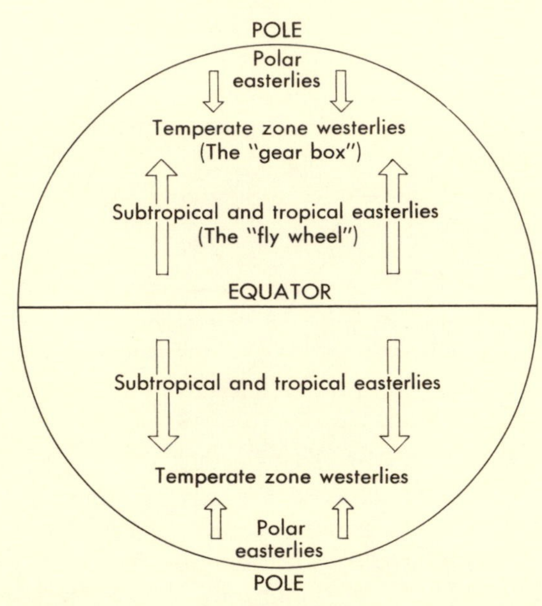

Fig. 17. The transfer of angular momentum in the general circulation (shown by arrows).

the Earth. The trade-wind belt, in this respect, is like a great fly-wheel into which there is an input of westerly angular momentum. A belt of westerly winds, by the same mechanism, loses westerly angular momentum to the Earth. If the circulation is to be maintained, there must be a continuous transport of angular momentum from the tropics into the belt of the westerlies, and in recent years there have been many efforts to find out if and how this happens.

The analysis (which is too mathematical to be given here)

shows that such transport can take place in two ways. First, any steady drift of air along the meridians (that is, northward or southward) from the tropics must carry with it some angular momentum. It is not easy to detect steady meridional motion from the observed winds, because it is so small, but it now seems to be generally accepted that there is a significant meridional component, especially in the upper air of the lower latitudes. The second way in which momentum can be transported is more subtle and probably more important. It was suggested by Sir Harold Jeffreys in 1926 that the characteristic large-scale disturbances of the mid-latitudes act as "gears" which transmit angular momentum from the tropical flywheel into the westerlies.

The mechanism proposed by Jeffreys is similar to that of the transfer of momentum by small-scale eddies in what is called *turbulent flow*. This is dealt with in detail in Chapters 6 and 7. The large amount of information on the upper winds of the globe that has become available since World War II has enabled meteorologists to compute the magnitude of the transfer from the observations. The result is encouraging. The required amount of transfer can be accounted for by the "gear" concept, with the trade-wind belts as the main source, and the westerlies the main sink, of angular momentum. Further, it has been shown by the Norwegian meteorologist J. Bjerknes and others that the pattern of easterly-westerly-easterly wind belts over the globe can exist indefinitely only if the trough lines of the great pressure waves and the axes of the cyclonic disturbances are orientated in the correct sense. Meteorologists may now claim not only to have located the main parts of the atmospheric engine but also to have a tolerably satisfactory idea of how it is designed. An interesting fact revealed by the investigations is that without the poleward transfer of angular momentum to replace frictional losses, the westerly winds of the mid-latitudes would die away in about a fortnight.

The problem of the general circulation is now one of the main preoccupations of meteorologists, and some notable advances have been made in the past decade. The investigations have been both mathematical and experimental. Work in fluid-motion laboratories with rotating bowls of water heated at the circumference and cooled at the center ("dishpan experiments") represent an attempt to imitate the atmosphere with its heat source in the tropics and its cold sink in the polar regions. As a result we now have a better idea of the way in which long waves and cyclones form in the real atmosphere. The mathematical experiments have been done with high-speed computers. The most imaginative work in this field is that of the American meteorologist N. A. Phillips, who set up a "model atmosphere" by deriving certain equations which represent, approximately, an atmosphere heated to the south and cooled to the north on a rotating hemisphere. The atmosphere was supposed to be at rest initially and (after many hours of computation on a very fast machine, corresponding to an interval of about five months in natural time) the temperature inequality created a circulation which had a slight resemblance to that actually found on the Earth. At this stage the motion was disturbed mathematically by putting in random numbers. Within a relatively short time (corresponding to some weeks of real time) a new circulation developed which not only bore a closer resemblance to the actual circulation but also had many features, such as long waves and closed circulations, that are found in the real atmosphere.

This experiment is significant in two ways. First, it shows that present ideas concerning the genesis and maintenance of the general circulation, crude though they may be, are essentially correct. Secondly, Phillips' work holds out the hope that someday it may be possible to apply similar methods to forecast changes in the circulation for relatively long periods—say several weeks—ahead. This cannot be done reliably today, but the development of a true system of long-range forecasting would be an event of great economic importance for the world as a whole.

The Jet Streams of the General Circulation

About 1946 a discovery was made at the University of Chicago which in some respects was as striking as that of the stratosphere. When, for the first time, sequences of upper-air charts were drawn to cover the whole of the Northern Hemisphere, it was found that there are clearly defined "rivers" in the high atmosphere, flowing at very high speeds. The name *jet stream* was coined for these narrow sinuous currents, some of which are shown in Figs. 18 and 19.

At about 18,000 feet above sea level Fig. 18 shows that there is a rather narrow circumpolar westerly current in which the wind

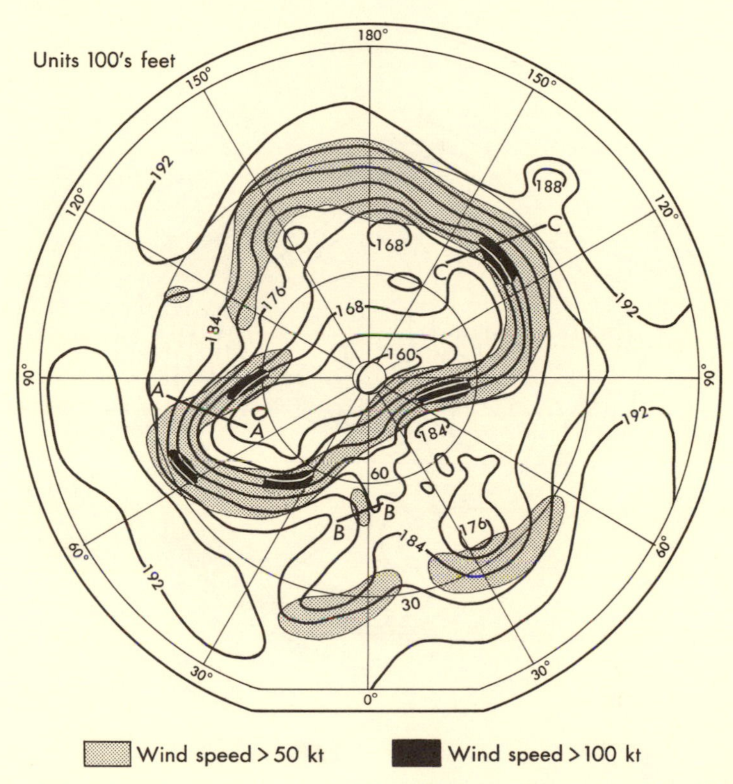

Fig. 18. 500 mb contours at 0300 GMT on December 19, 1953. (After Sawyer.)

speed is greater than 50 knots[9] and in some parts reaches 100 knots. In Fig. 19 the strong westerly current is sinuous and narrow but more complex, and again 50 knots is frequently exceeded

[9] In professional meteorology the unit of wind speed is the knot, equal to 1 nautical mile an hour. In scientific studies the metric system is more common. Approximately, 2 knots = 1 meter a second.

and there are extensive areas with average winds over 100 knots. It is now generally accepted by meteorologists that there are two main narrow currents moving at high speed in the high atmosphere, the *subtropical jet stream,* which is found between 30,000

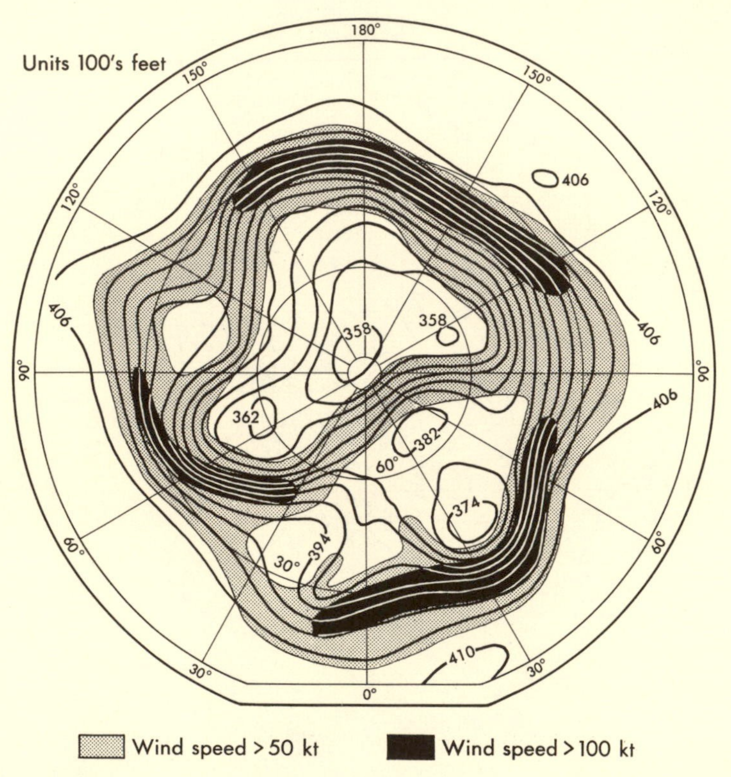

Fig. 19. 200 mb contours at 0300 GMT on December 19, 1953. (After Sawyer.)

and 40,000 feet above sea level in the "horse latitudes" (between 30 and 35°), and the *circumpolar jet stream,* which is found much farther away from the equator, as Fig. 18 shows. The circumpolar jet stream seems to be more subject to distortions and disruptions than the subtropical jet, which is almost certainly a consequence of the Hadley-cell type of meridional circulation which exists in the low latitudes. The meridional motion, although less in evidence than the zonal motion, is clearly an exceedingly important ingredient in the general circulation. One meteorologist has indeed gone so far as to regard the zonal motion as an accident caused by the rotation of the Earth and the meridional motion as the prime mover.

The problem of the general circulation is still far from complete solution, but it may well be that one of the most serious handicaps, that of incomplete data, will be greatly reduced in the not too distant future. In the first place, the International Geophysical Year provided an opportunity for meteorologists to measure the characteristics of the atmosphere on a scale and with a detail never attempted before. Secondly, the development of artificial satellites will furnish a means whereby the motion of the air over the whole globe, as shown by clouds, can be seen for the first time. As a result of these advances it may be that the picture of the circulation sketched above will need to be changed in some respects, but time alone can say if this will happen. The problem of long-range forecasting, of the prediction of weather on a monthly or even a seasonal basis, is not likely to be solved until more is known about the general circulation, and this is one of the reasons that meteorologists regard the study of the major dynamical systems of the atmosphere and their interactions as a matter of urgency.

CHAPTER 3

THE PHYSICS OF CLOUDS AND RAIN

The great wind systems that make up the general circulation control, very largely, climate. We have to consider next the smaller but still large systems that determine weather, such as the depressions and anticyclones of the mid-latitudes. Before we can examine the anatomy of such systems with profit it is necessary to understand something of the physical processes that take place in them—the physiology, so to speak, of the organism. In particular, it is essential that there be a clear conception of the fundamental process of condensation, which is not as simple as may appear at first glance.

Clouds

The study of clouds may take one of two forms. We may be interested chiefly in their structure as static entities, in the nature and mode of formation of the drops and particles of which they are formed. This aspect of meteorology is referred to as the *microphysics of clouds*. On the other hand, we may regard a cloud as a dynamic entity which forms, grows, and finally dissipates as the result of large-scale movements of the atmosphere. This is the *macrophysical* aspect, which looms large in dynamical meteorology and forecasting.

A cloud is an aggregate of particles, chiefly of water in either liquid or frozen state, which is formed by the chilling of moist air to a temperature below its dew point (p. 17). The chilling may be brought about by the lifting and consequent expansion of the air or by the cooling of air near the ground by radiational loss of heat. In the lifting process, cooling is nearly adiabatic, so that until the dew point is reached the temperature falls at the dry-adiabatic lapse rate. The latent heat liberated by condensation reduces the rate of cooling to the saturated adiabatic lapse rate, which except at very low temperatures is about half the dry-adiabatic rate.

In professional meteorology clouds are classified by their appearance, the international system used today being much the same as that originally proposed by the Quaker scientist Luke Howard in 1803. Howard based his nomenclature on the three Latin words *cirrus* (hair), *cumulus* (pile or heap), and *stratus* (layer), but since his day many other names have been introduced. Cirrus clouds are white, with a threadlike structure; they are invariably found at great heights. The cumulus type of cloud is easily recognized by its characteristic tower or cauliflower appearance and its flat base. Stratus is the name given to extensive layers of cloud with little or no detail. For the purpose of the present account it is not necessary to go into greater elaboration, and the reader who is interested in clouds for their beauty is advised to look into the International Cloud Atlas published by the World Meteorological Organization, where he will find much to delight his eye.

In scientific papers the Howard classification is now less used than before, and the research worker tends increasingly to refer to clouds in terms of the physical processes by which they are produced, instead of by their appearance. Cumuliform clouds are now often called *convection clouds* because they owe their formation to strong upward currents. Layer clouds, on the other hand, depend largely upon horizontal motions, and three types may be distinguished: those which occur as a result of air being forced to cross mountain ranges (*orographic clouds*); those which are formed by the large-scale mixing of air masses of different temperatures and water-vapor contents; finally, those formed by the slow widespread ascent of air consequent upon certain pressure patterns. The last-named are responsible for much of the rain that falls in the temperate zones of the Earth.

Orographic clouds usually are lenticular (lens shaped) or display undulations, and for this reason are often called *wave clouds*.

The disturbance set up by a mountain range causes waves to form over and in the lee of the hills, and such waves may persist for considerable distances downwind of the summit. Even quite low hills can cause thin wave clouds, and on occasion it is possible to see at great heights (up to about 20 miles) over the mountains of the northern latitudes stratospheric wave clouds of the type known as "mother-of-pearl" because of their brilliant iridescent colors.

Layer clouds caused by the large-scale mixing of air masses include fogs, such as those of the Grand Banks off Newfoundland, which are the result of warm moist air being brought into contact with air cooled by contact with a cold surface. The most common form of layer cloud is, however, that which accompanies depressions, especially in the vicinity of fronts. Such cloud systems may cover thousands of square miles, and they carry an enormous weight of water.

In cumuliform clouds, which are the result of strong convection, there is evidence that the ascent of air takes the form of the rise of discrete "bubbles," the process resembling the boiling of a liquid. Cumulus clouds are continually changing in shape, because of the incessant condensation and evaporation. In thundery conditions the updraft may vary from 15 to 100 feet a second, and drops can be swept upward from the base of the cloud to its top thousands of feet above in a matter of 20 minutes or less. Flying in the vicinity of large cumulus clouds can be uncomfortable, and for this reason modern aircraft are often fitted with radar detectors which give the pilot good warning of the highly disturbed areas.

Radar has proved to be a most useful tool for the study of clouds. Centimetric radio waves are well reflected by drops and ice particles of diameter exceeding about a millimeter, and as a result it is possible to examine in fair detail clouds extending over large areas. Such studies show that few clouds are uniform in structure but are generally composed of many "cells," whose period of life as recognizable entities may be anything from 20 minutes to several hours. This is especially so in convection shower clouds and thunderstorms, but the same conclusion also holds for what is recorded by the weather observer as "steady rain" over a large area. A lapse-time cine film of the radar record of what appeared to be a night of steady rain over the whole of the London area showed a continuously changing pattern of precipitation, with many gaps, and there is no reason to think that this was an exceptional case.

The liquid water content of clouds is highly variable and concentrations up to 5 grams per cubic meter have been recorded. The total weight of water in even a moderate-sized cumulus cloud, which may be made up of drops, ice particles and snowflakes, is enormous, probably several hundred thousand tons at least. The first question that arises is therefore how such great weights can remain aloft for long periods. The drops clearly cannot "float" because water is about 800 times as dense as air. The answer lies in the minute size of a cloud particle, which rarely exceeds a few hundredths of a millimeter. When such a body begins to fall under gravity through the atmosphere it sets up an air resistance proportional to its speed. This resistance becomes equal to the weight of the drop (which is very small) at a very low speed, and once this condition is reached there is no longer a net force to cause acceleration. The limiting speed of fall, called the *terminal velocity,* is so small for cloud drops that the downward motion is virtually imperceptible and the cloud appears to float in the sky.

It is thus a simple matter to explain how clouds can remain aloft indefinitely. The real difficulty for the meteorologist is to explain how the water ever gets down—in other words, how are large raindrops formed from small cloud drops? It is only within the past twenty years that convincing answers have been given, but to comprehend the theories we must first inquire more closely into the processes by which the water vapor of the air is condensed into cloud drops.

Condensation in the Atmosphere

In Chapter 2 we touched lightly upon the physical facts about vapor in the air, but for the present purpose it is essential to look at the matter in a more exact fashion. The molecules in a volatile liquid are in a state of incessant activity which causes some of them to escape from the surface of the liquid into the space above, which may or may not contain another gas. If the liquid partly fills a closed vessel, some of the molecules that escape are recaptured by the liquid as they strike its surface, but eventually a state of dynamic equilibrium is reached in which the number of molecules that escape is exactly equal to the number that are recaptured. The vapor in the closed space above the liquid is then said to be *saturated,* and its pressure has a definite value at any given temperature. The same arguments apply when the liquid is frozen, but there are two sets of values for the saturation vapor pressure of water, corresponding to equilibrium over a plane surface of water, and of ice, respectively. The values differ, with the pressure over water significantly greater than that over ice.[1] This property of water is of great importance in cloud physics.

If the vapor is allowed to diffuse freely into space by leaving the mouth of the vessel open, the number of molecules leaving the surface is unaltered but less of them return. The vapor is then said to be *unsaturated* and evaporation continues until all the liquid is converted to gas. It is also possible for a vapor to be *supersaturated.* This state can be attained momentarily in the laboratory by subjecting saturated vapor to a sudden large expansion, as in the famous Wilson "cloud chamber" experiment.

The definitions given above make no reference to the presence of gases other than the vapor of the liquid in the space above the surface. In meteorology it is customary to refer to "saturated air," but this, strictly, is inaccurate. It is the water vapor, not the air, that is saturated, and it is misleading to speak of the atmosphere as if it were a kind of sponge which, at a given temperature, can hold a certain amount of water vapor and no more. But the usage is sanctioned by custom, and it would be pedantic always to insist on the accurate phraseology. Here we shall follow fashion and use the terms "saturated" and "unsaturated" air freely when there is no risk of ambiguity.

In the supersaturated state a vapor is likely to condense spontaneously on any solid or liquid surface, but the process by which small drops are formed in the atmosphere is by no means as straightforward. If a volume of moist air is carefully cleaned by repeated filtration it is possible for the water vapor it contains to

[1] See p. 65 for the actual values.

sustain high supersaturations, up to several hundred per cent, without drops appearing in the body of the gas. In the atmosphere supersaturation rarely exceeds a fraction of 1 per cent, so that one is led to conclude that the specks of dust removed from atmospheric air by filtration in the laboratory experiment must play an important part in the formation of natural clouds.

It is now recognized that when water vapor condenses to form clouds it always does so on myriads of tiny particles called *condensation nuclei.* The essential part played by such nuclei in promoting condensation was first demonstrated by the French chemist M. Coulier in 1875 and later, in 1880, by the Scottish physicist J. Aitken, who showed that mists formed in slightly supersaturated air are made denser, or thinner, by the introduction, or removal, of combustion products. The atmosphere always contains an abundance of condensation nuclei, and this accounts for the fact that it is never more than marginally supersaturated. Condensation occurs in the atmosphere as soon as the vapor pressure exceeds the saturation value because there are plenty of nuclei present.

Despite intensive research, the nature and exact mode of action of atmospheric condensation nuclei have not been fully elucidated. It has been established that the nuclei vary considerably in size, ranging in diameter from a few millionths to a few thousandths of a centimeter, and that their concentration (number of nuclei per cubic centimeter) also varies considerably from place to place. As would be expected, the lowest concentrations, less than a hundred per cubic centimeter, are found over the oceans and at great heights. In the smoky air of settled areas concentrations as high as several million per cubic centimeter have been measured. Nuclei may be hygroscopic, like salt, or nonhygroscopic, like carbon. They may be formed by mechanical disruption, such as the erosion of rocks and soils, or by the breakup of sea spray by the wind, or they may come from forest fires, volcanic eruptions and industrial processes. Domestic chimneys undoubtedly make a significant contribution to the total nucleus content of the atmosphere.

The essential role of nuclei in condensing the water vapor of the atmosphere, and the fact that so many nuclei are produced by fires, raises the important question whether man-made atmospheric pollution is to be regarded as a primary cause of fog. The answer appears to be that, although smoke makes fog denser and possibly more persistent, it is not a direct cause of fog, even in great cities. In a country such as Britain the prohibition of open fires would not get rid of the ever-recurring nuisance of winter fogs. There are always enough natural nuclei in the air to cause a fog whenever the conditions of temperature and humidity are favorable. There are many good reasons why atmospheric pollution should be reduced, but the elimination of fog is not one of them.

One other property of water is important in a discussion of clouds, namely, its ability to *supercool* or remain liquid at temperatures below 0° C (32° F). There is no definite freezing point of water, but only a fixed melting point of ice. Water as supplied for drinking, or as found in rivers, lakes and ponds, contains many tiny particles which act as *freezing nuclei* and ensure that ice is formed at temperatures just below 0° C, but cloud drops are less likely to contain such nuclei and can remain liquid down to −40° C, when all water drops freeze. A few cloud drops freeze spontaneously at temperatures between 0 and −10° C, and slightly more between −10° and −30° C, and a high cloud nearly always contains vast numbers of supercooled drops, unless the air temperature is below −40° C.

We have thus to take into account both condensation nuclei and freezing nuclei in the physics of clouds. It has been found that many substances act as freezing nuclei when they are in a finely divided form. Ice is very effective, and a supercooled drop freezes at once if it comes in contact with an ice particle. Silver iodide also is remarkably effective, causing some drops to freeze at temperatures as high as −5° C and all drops to freeze below −10° C. Other substances which on coming into contact with a supercooled drop cause it to solidify are loam, clay, ash and sand; they generally act between −15° and −25° C and thus are less effective than silver iodide. Pellets of solid carbon dioxide ("dry ice") act because of their very low temperature (about −78° C) and any substance below −40° C behaves similarly. The physics of the process is not yet fully understood.

Rain, Snow and Hail

One of the central problems of cloud physics is to ascertain how the minute cloud drops grow or combine to form raindrops, snowflakes and hailstones. The smallest drops that reach the ground do so as *drizzle;* their diameters are about a tenth of a millimeter. *Raindrops* have diameters up to about 7 millimeters, and *hailstones* normally up to about 10 centimeters. Even small raindrops are equivalent in volume to the aggregation of a million or more cloud drops, and the hard core of the problem is to explain how this takes place in some clouds and not in others.

An obvious suggestion for the formation of rain is that cloud drops grow into raindrops by the continuous acquisition of water in a slightly supersaturated atmosphere. Unfortunately, this simple explanation does not bear detailed examination. The rate of growth of a drop depends upon two factors: the diffusion of water vapor from the air to the drop and the conduction of the latent heat of condensation from the drop to the air. If a reasonable degree of supersaturation is assumed (say 0.05 per cent), it is possible to make numerical estimates of the time taken for a typical cloud drop to grow to any prescribed size. The results show that the rate of growth is rapid at first but soon becomes very slow. The conclusion is that, although drizzle drops can be formed from cloud drops by condensation and growth in a thick cloud, the process takes several hours and to reach raindrop size would require an impossibly long time. The largest drops that are likely to form by steady condensation have diameters not exceeding a few hundredths of a millimeter and would almost certainly evaporate in the warmer surface layers before reaching the ground. The spontaneous coalescence of small drops to form large drops also must be regarded as unlikely because of the relatively great distances, up to 50 diameters, between drops in a cloud, and unrealistic because clouds often persist for days.

We must therefore look for a selective process, such as the amalgamation by collision of drops moving at different speeds. Conditions favoring such a process are found in large cumulus clouds, especially in the tropics, where the ambient temperature is above 0° C. Observations made in these clouds show that, although most of the drops are very small, there are always present some "large" drops, several hundredths of a millimeter in diameter, which are probably formed from large hygroscopic nuclei. Such drops, when they are swept upward by the strong convection currents in the cloud, grow as the result of collisions with the small drops. If the cloud is deep and the water content high, the big drops, when they approach the top of the cloud, become too heavy to be sustained by the updraft. They fall back, growing even larger by more collisions, and eventually fall out of the cloud. Some drops reach a size at which they are unstable and break up under the stresses imposed by their motion, thus providing a fresh supply of "large" drops. In this way a chain reaction is set up and a heavy shower results.

The collision process is believed to be responsible for the typical tropical shower, since in these conditions convection is strong and the whole of the cloud is above 0° C. The process by which rain is formed from deep-layer clouds, with many supercooled drops and a few ice crystals in their upper layers, is quite different. This process, first worked out in detail by the Swedish meteorologist T. Bergeron, depends upon the difference between the saturation vapor pressures of water over water and over ice, respectively. The following values show this difference as a function of temperature:

Saturation vapor pressure over water and over ice

Temperature (°C)	0	−10	−20	−30	−40
Sat. v.p. over water (mb)	6.1	2.86	1.25	0.51	0.19
Sat. v.p. over ice (mb)	6.1	2.60	1.03	0.38	0.13

In a cloud composed of large numbers of supercooled water drops and relatively few ice crystals, air saturated with respect to water is supersaturated with respect to ice. As a result there is a preferential acquisition of water by the ice particles, which soon grow into star-shaped crystals large enough to fall through the drops. In the fall tiny fragments of ice break off from the fragile arms of the crystals to form new rapidly growing crystals, so that the process is self-sustaining. In this way a cloud which originally had a preponderance of supercooled drops can change fairly quickly into one almost entirely made up of ice crystals. The crystals interlock and form snowflakes, which are large enough to float downward. On a wintry day, with the air temperature everywhere below 0° C, the final result is a snowstorm, but in warmer weather the snowflakes melt in the surface layers and become raindrops. Most of the steady rain from the deep-layer clouds associated with mid-latitude depressions begins as ice and snow in this way.

Hail occurs in both winter and summer and is frequently found in thunderstorms, which points to convection as an essential element in its formation. The Bergeron mechanism requires the continuous renewal of the supply of ice particles from star-shaped crystals which grow mainly as the result of the diffusion of water vapor, and not by collision. This implies that the water content of the cloud must not be too high. In convection clouds, with vigorous vertical currents and plenty of drops, collisions occur frequently, with the result that the ice particles become roughly spherical aggregates with many air spaces. Such pellets are called *graupel* or *soft hail.* It is more difficult to explain the formation of the large hard *hailstone* with its onionlike structure of alternate clear and opaque layers of ice. The most popular theory is that the opaque layer is formed in the relatively cold low-water-content layers of the cloud, where supercooled drops freeze on capture, and that the transparent layers are formed from a liquid coat, acquired in the dense warmer parts of the cloud, which soon becomes frozen. The existence of more than one skin is attributed to the fact that the hailstone is swept upward and downward many times by the powerful vertical currents before it grows so massive that it must fall to earth.

In the United States a survey shows that most of the hail that falls is between 1 and 2 centimeters in diameter. The largest hailstone observed in the United States up to 1957 was 13.8 centimeters (5.4 inches) in diameter and weighed 1½ pounds. This stone fell in Nebraska. A hailstone weighing 7½ pounds has been reported from Hyderabad, India.

To sum up we may say that the formation of rain, snow and hail from vapor is now becoming fairly well understood by meteorologists. Figure 20, reproduced from Dr. B. J. Mason's treatise *The Physics of Clouds,*[2] shows schematically the possible ways in

[2] Oxford University Press, 1957.

which cloud drops can become large enough to fall to the surface and so complete the water cycle which began with evaporation by the heat of the Sun.

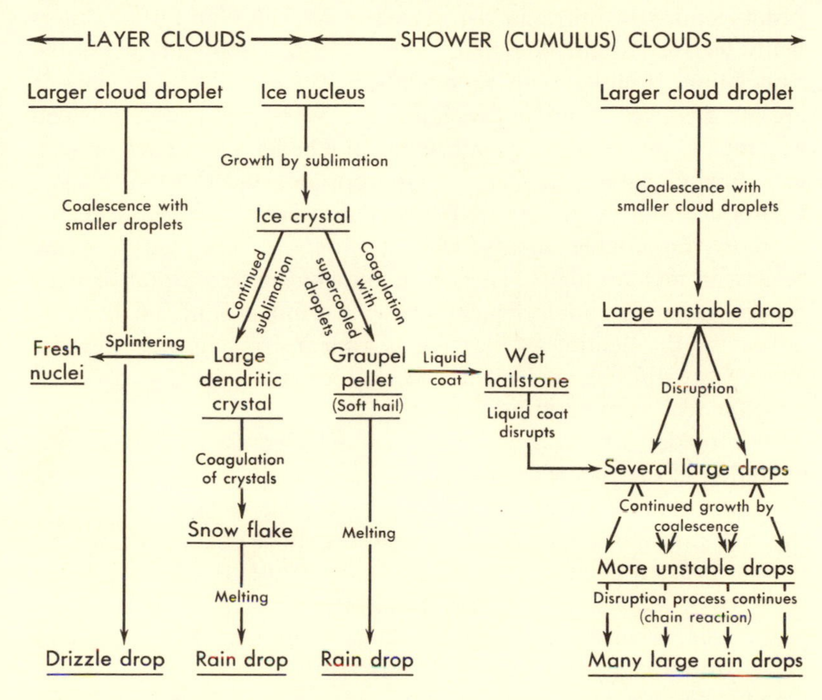

Fig. 20. A schematic diagram showing natural precipitation mechanisms. (Mason.)

Man-Made Rain

Man has always depended for his existence upon an abundant supply of rain. It is therefore natural that attempts to induce rainfall have been made since the dawn of history, at first by magical and religious rites and in the past hundred years by more direct means, but it is only within the past thirty years that our knowledge of the physical processes involved in the natural production of rain and snow has been adequate to give scientific guidance to the rainmakers. Even now it is difficult to judge whether success can be attained on any worth-while scale by means that are attractive economically.

There is an old and widespread belief that very loud sounds, such as those of gunfire, cause rain. Many of the great battles of history ended in rain and the period of World War I was abnormally wet in Britain, but there is no scientific foundation for the association of noise with precipitation. Even the loudest sounds are mere ripples compared with the large surges of pressure which accompany depressions. Nevertheless, the belief was strong enough to induce Congress, in 1891, to appropriate $9,000 for a project to produce rain by gunfire in Texas. The results were inconclusive. A more reasonable theory was that of the nineteenth-century American meteorologist James P. Espy who, realizing that upward motion is an essential element in the natural production of rain, advocated the lighting of bonfires as a means of relieving drought. What Espy did not realize was the scale of the processes involved, which made his method ineffective.

The reader who has followed the description of natural precipitation in the preceding pages will be in a position to appreciate that the artificial introduction of very cold substances in a finely divided form into a supercooled cloud has more chance of success. Such a process is called "seeding." In 1930, before Bergeron had made clear the role of freezing nuclei in the initiation of precipitation in supercooled layer clouds, a Dutchman named Veraart used solid carbon dioxide ("dry ice") and supercooled water to seed clouds. He claimed to have produced rain, but his results were received with skepticism by his fellow countrymen because of the seeming lack of theoretical support for his methods. Without a clear understanding of the chain reaction caused by the splintering of ice crystals it is certainly difficult to believe that the sprinkling of a few pounds of dry ice on the top of a cloud can possibly cause the release of thousands of tons of water. Veraart's work came too early for its potentialities to be appreciated.

The first significant step toward weather modification came with the discovery by Dr. Vincent Schaeffer in 1946 that a few particles of solid carbon dioxide dropped into a vessel containing a mist of supercooled drops produces millions of ice crystals. (The experiment is easily done in a deep-freeze cabinet, when the presence of ice crystals is clearly shown by the sparkling appearance of the mist in the beam of light from a torch.) Schaeffer soon followed up this discovery by a spectacular demonstration of the changes caused in a natural cloud of supercooled drops by pellets of dry ice dropped by aircraft. Large volumes of the clouds were converted to ice crystals and in some instances precipitation was observed. Here was proof that the Bergeron mechanism works in the atmosphere and that, in theory at least, rainmaking is possible in suitable conditions.

Soon afterward Dr. Bernard Vonnegut discovered the remarkable ice-nucleating properties of silver iodide. Vonnegut was led to examine this apparently unpromising substance, which is not a normal constituent of the atmosphere, because its crystal structure is like that of ice. Silver iodide is not prohibitively expensive and is easily dispersed in a finely divided form, and it seemed that the way to the artificial inducement or augmentation of rain had opened up in a remarkable fashion. Several full-scale projects were soon under way. "Project Cirrus" was conducted by the General Electric Company, with the support of the armed forces, and nearly two hundred trials were made by this group, mainly of the seeding of clouds with dry ice and silver iodide. Another major investigation, the "Cloud Physics Project," was conducted jointly by the U.S. Weather Bureau, the U.S. Air Force, and the National Advisory Committee for Aeronautics.

There was, however, considerable other activity which often lacked the scientific control and impartiality of the official projects. The seeding of clouds by aircraft is expensive and difficult to extend to cover large areas, and as a method of producing rain to save a crop is hardly likely to recommend itself to the private user of limited means. Vonnegut's discovery of the nucleating properties of silver iodide suggested a far easier and cheaper way of seeding clouds. It is a simple matter to produce copious clouds of silver iodide "smoke" (that is, masses of airborne particles) from ground "generators" like those used to produce screening smokes in war. The problem is, in fact, easier than that which confronted the armament scientist, for adequate concentrations of silver iodide particles can be produced by sublimation, using braziers of coke impregnated with silver iodide, or apparatus working on the blowlamp principle with silver iodide dissolved in acetone. A line of such generators, widely spaced on a front extending many miles across wind and sited some miles upwind of the area on which it is desired that rain should fall, produces plumes of particles which, in theory at least, diffuse upward into the supercooled regions of the natural clouds and initiate the Bergeron process. In this way it was hoped that large volumes of recalcitrant clouds could be seeded cheaply, for only unskilled labor is required to install and work the generators. The method proved irresistibly attractive to many commercial operators, some of whom rushed into the field with an inadequate appreciation of the many uncertainties involved both in operation and in the assessment of the results, and some very extravagant claims of success were made in the early years. As a result there have been acrimonious disputes between the professional meteorologists and the commercial operators, and it has been difficult for the outsider to form an objective view of the matter.

The initiation or augmentation of rain by silver iodide particles produced at ground level is not straightforward. A cloud of small particles formed on the ground is carried by the wind and dispersed in the lower layers of the atmosphere by eddies. Near the surface the rate of upward diffusion is high, and in daytime the particles soon reach heights of several hundred feet. Above this level there are fewer eddies to scatter the smoke and the rate of upward diffusion falls off sharply with height. If there is a strong temperature inversion in any layer, the upward motion of the particles is reduced to zero and the smoke travels horizontally. It is therefore difficult to be sure that nucleating material released at ground level will reach the supercooled layers of the cloud in appreciable concentrations. A further handicap is that silver iodide loses its nucleating power by prolonged exposure to sunlight.

The assessment of the results is very difficult. Even when rain falls during or soon after the release of the silver iodide, it is impossible to decide immediately whether or not the trial has been successful. In laboratory physics a result is rarely open to serious doubt, for an experiment can be repeated any number of times in the same conditions. This does not hold in meteorology, for no two states of the atmosphere are exactly identical. There is always the possibility that rain which fell during or after seeding would have fallen in any event, irrespective of the intervention of man, for seeding can have no chance of success unless nature has already set the scene for the natural production of rain by bringing up supercooled clouds in sufficient depth. There is no way of distinguishing "artificial" from "natural" rain on the ground, and on any single occasion it is never possible to say with certainty that seeding induced or even augmented rain. A statistical examination of the results of a long series of trials offers the only hope of reaching a true verdict.

Rainfall is the most variable of all the meteorological elements, both in space and in time. Two rain gauges placed a mile or more apart in level country are likely to show substantial differences over short periods, and long records of rainfall measured at a fixed spot show large irregularities. It is now known that the effects of seeding are not spectacular and amount at the most to increases not greater than the normal deviations from the average. The task of the scientist in assessing the results of seeding trials is therefore to decide whether small fluctuations in a curve which is naturally highly irregular can be attributed to intervention by man. This calls for delicate and elaborate statistical analysis.

There are two ways in which rainmaking operations can be tested statistically. One is to select an area for which reliable rainfall records have been maintained over a long period, not less than ten years. Trials with ground generators must then be conducted for several years, for single trials or short runs are meaningless. If it can be shown that during the period of the trials the rainfall in the area was *significantly*[3] greater than the long-term average, it may fairly be said that the seeding had the desired effect. The second method is to take two adjacent areas for which there is a close relation between precipitation, established by comparison of measurements taken over long periods. One area is seeded and the other treated as the control. Again, single trials or short runs are valueless, and a statistically significant difference must be established

[3] The statistician decides whether the difference between two variable quantities is "significant" by calculating the odds against the result being due to chance, i.e., being part of the natural fluctuations. The method is objective, but care has to be taken in its application.

between the areas before it is safe to assert that rainfall is affected by seeding. There is little doubt that the second method is the safer to use because of the well-known tendency for wet years to come in groups.

Tests of this kind were used by the Advisory Committee on Weather Control which published its Final Report at the end of 1957. The main conclusions reached are as follows:

1. The statistical procedures employed indicated that the seeding of winter-type storm clouds in mountainous areas in western United States produced an average increase in precipitation of 10 to 15 per cent from seeded storms, with heavy odds that this was not the result of natural variations in the amount of rainfall.
2. In nonmountainous areas, the same statistical procedures did not detect any increase in precipitation that could be attributed to cloud seeding. This does not mean that effects may not have been produced. The greater variability of rainfall patterns in nonmountainous areas made the techniques less sensitive for picking up small changes that may have occurred there than when applied to mountainous regions.
3. No evidence was found in the evaluation of any project which was intended to increase precipitation that cloud seeding had produced a detectable negative effect on precipitation.
4. Available hail frequency data were completely inadequate for evaluation purposes and no conclusions as to the effectiveness of hail suppression projects could be reached.

The positive result claimed for mountainous areas is explained by the fact that the upcurrents created by a long range of hills rapidly carry the nuclei into the supercooled layers of the cloud in high concentrations. Seeding thus has its greatest chance of success when the generators are placed on the windward slopes of high mountains in winter, but it is only fair to say that not all meteorologists accept even this result as proved beyond doubt.

The guarded language of the committee's report contrasts strongly with the statements of some of the commercial operators, who in the early stages claimed increases in rainfall of 100 per cent and more. Whether or not man can modify this vital element of the weather is still not decided, but it seems unlikely that a farmer whose fields are suffering from lack of rain can remedy matters simply by arranging for silver iodide smoke to be generated on his property. As regards the problem of artificial precipitation in general, a poll of informed opinion taken throughout the world might well result in the old Scottish verdict of "not proven." In the most pressing problem of all, the alleviation of a long drought, seeding offers no hope. Conditions must be such that rain is falling or is likely to fall naturally before there is any chance of success with artificial nucleation.

Attempts have also been made to induce showers from warm cumulus clouds by the introduction of large drops to initiate the collision process. Such drops can be formed by condensation on large hygroscopic particles, and common salt has been dispersed from rockets to this end. So far the trials (mainly in tropical areas) have been too few for any conclusion to be drawn.

CHAPTER 4

WEATHER-PRODUCING SYSTEMS

In Chapter 2 we discussed the basic pattern of winds over the globe. We have now to consider the smaller but still large systems associated more with weather than with climate. Such systems, the cyclones (lows or depressions) and anticyclones (highs) which enter so largely into the daily forecasts, are complex and their exact, detailed analysis is beyond the present powers of applied mathematics. It is possible, however, to gain considerable insight into the behavior of these systems by restricting attention to their main physical features. The cyclones and anticyclones of dynamical meteorology are idealizations of the complicated patterns which nature creates, and theory can give no more than an imperfect account of what actually occurs. The form of the idealization is suggested, very largely, by two fundamental concepts of classical hydrodynamics—*vorticity* and *divergence*—which have long been recognized by meteorologists as especially significant in the problems of the atmosphere.

Vorticity and Divergence

When a mathematical physicist talks of a *fluid* he has in mind a medium which differs in many ways from the real fluids of the experimental physicist. The mathematician claims that he is at liberty to divide a fluid into infinitesimally small volumes without sacrificing any of the bulk properties of fluids, such as density or temperature, in the process. The experimental physicist is quite certain that such a process would result in nothing more than empty spaces, devoid alike of mass and intellectual appeal. But if we are to apply the calculus to what is really a collection of discrete molecules we must be prepared to face logical difficulties of this kind. The fluid of the mathematician is a hypothetical continuous medium to which certain operations of mathematics are applied freely in the hope that the result will throw light on the behavior of real fluids. The assumption of continuity as an approximation to reality is amply justified in the lower parts of the atmosphere, because here the average distance between molecules is small compared with the scale of the systems studied, and the time between molecular collisions is small compared with the time taken to make an observation, but at very great heights the concept of a continuous fluid no longer holds.

The word "vorticity" conjures up visions of whirlpools, but in fluid mechanics it is simply the angular velocity of the infinitesimal elements, or fluid particles, into which the mathematician supposes that the fluid is divided.[1] When such particles move they may be *translated, deformed* and *rotated,* and in general all three types of motion are present.

Here we are interested only in rotation.[2] It is clearly impossible to examine directly how the little blobs of air are rotating, and if vorticity is to be used in real problems, it must be expressed in terms of quantities that can be measured. To specify vorticity in a quantitative manner we must attribute to it *magnitude, direction* and *sense.* (In mathematical terms, this means defining vorticity as a *vector.*) This is not difficult. Magnitude is defined as twice the angular velocity, the number of degrees (or radians) that the particle turns through every second. Direction is specified by taking note of the axis about which the particle rotates. A particle rotating like a gramophone turntable, entirely in the horizontal plane, may equally be thought of as rotating about a vertical axis and one rotating in the vertical plane, like the wheels of an automobile, as spinning about a horizontal axis. The direction in space of the axis of rotation thus affords a means of defining the direction of vor-

[1] In practice it is more convenient to define vorticity as *twice* the angular velocity. This convention is followed in this book.

[2] The reader who would like to know more of the fundamentals of fluid motion will find an account in the writer's *Mathematics in Action,* Chapter 5 (Harper Torchbooks, The Science Library, 1960.)

ticity without ambiguity. Finally, to complete the specification, the sense of the rotation must be defined, and here the rotation of the Earth, which is always in the west-to-east sense, forms a useful reference. In meteorology a particle of air which, relative to the Earth, rotates in the same sense as the Earth does in space is said to have positive, or cyclonic, vorticity. Anticyclonic, or negative, vorticity denotes rotation relative to the Earth in the opposite sense.

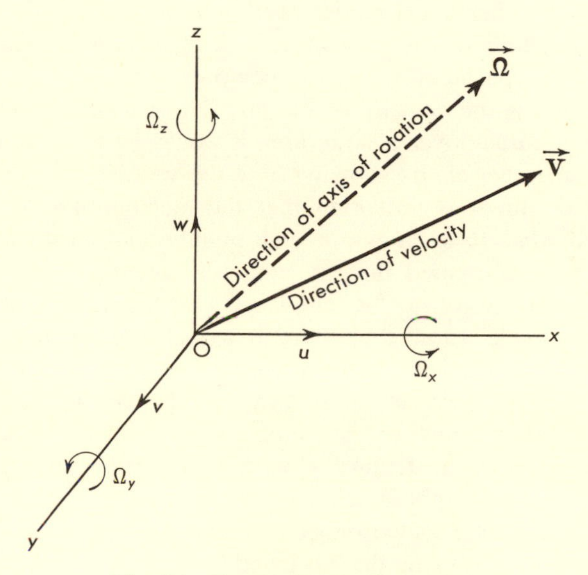

Fig. 21. Components of vorticity.

In Fig. 21, a particle is rotating about an axis in space. For the purposes of calculation it is convenient to put into this figure a frame of reference, specified by three mutually perpendicular axes *Ox* and *Oy* (horizontal) and *Oz* (vertical), just as we specify position on the Earth by latitude, longitude, and height above sea level. We can then imagine that the total rotation Ω is shared out between the three axes, just as the motion of an aircraft flying from America to Europe may be thought of as so many knots along a line of latitude plus another speed along a line of longitude together with its rate of climb in the vertical. We call Ω_x, Ω_y and Ω_z the *components* of vorticity along their respective axes, just as a velocity **V** is thought to be made up of components u (along Ox), v (along Oy) and w (along Oz).

It is then an easy matter to show that the components of vorticity are expressed mathematically by simple formulas which involve only the differences of the rates of change of the wind components along pairs of axes that are perpendicular to the directions of the components. The formulas are quoted in Appendix II, but the nonmathematical reader need note only that with their aid we can forget about the impossible task of measuring directly the rotation of air particles in order to compute vorticity. All that need be measured is the velocity of the air. If this could be done with sufficient accuracy over a large area it would be possible immediately to construct charts showing how the vorticity of the wind is distributed. But it is not often that the meteorologist has at his command wind observations of such quality and density that vorticity can be computed directly with real accuracy, and indirect methods often have to be employed. Such charts, when they have been made, show that there is a real relation between the distribution of vorticity and weather over large regions of the Earth's surface. This, in many ways, is the key to the modern mathematical approach to the problems of weather analysis.

One very important feature of vorticity in large-scale dynamical systems of the atmosphere helps to simplify matters for the meteorologist. In synoptic meteorology only the vertical component Ω_z, which measures spin in the horizontal plane, is significant. This is a consequence of the fact that the large-scale movements of the atmosphere are very nearly horizontal. (In the following pages the qualifying words "vertical component of" will often be omitted and "vorticity," when it is mentioned, must be understood to mean rotation in the horizontal plane.)

In small-scale atmospheric motions the vertical component of vorticity is not necessarily dominant, and rotation of the particles may be present in what may appear, at first sight, to be an irrotational current. In shallow layers near the ground the wind invariably increases with height, for the air in contact with the ground is always at rest. In certain conditions, which are considered in Chapter 6, the motion is smooth and the air particles glide over each other in a series of horizontal planes. The motion is like that of an array of horizontal roller bearings, with the upper layers rolling over the lower layers. The increase of the wind with height (the *wind shear,* in technical terms) means that a line of particles is rotated and the vorticity is simply Ω_y, with Ω_x and Ω_z both zero (Fig. 22).

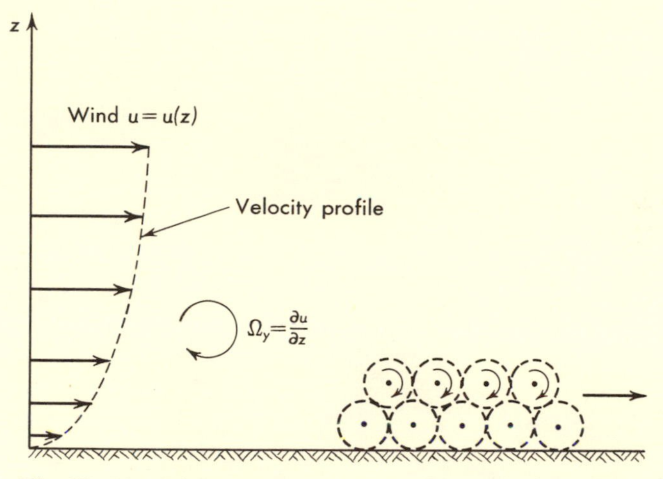

Fig. 22. Vorticity caused by wind shear near the ground.

Vorticity is found in large amounts only in relatively small regions of the atmosphere, for example, along boundaries between warm and cold air, and may be transferred from one part of the fluid to another, just as heat is transferred. The most important property of vorticity for large-scale systems is, however, that known to meteorologists as the *conservation of absolute vorticity.*

An observer in space, watching the movements of clouds on the Earth, would regard their rotation as made up of that of the air relative to the Earth plus the rotation of the Earth itself. This quantity is called the *absolute vorticity.* If we consider a particle of air at rest relative to the Earth, it is clear that when it is over a pole it is rotating entirely in the horizontal plane at the angular speed of the Earth (ω). The vertical component of the vorticity of the particle is then 2ω. On the other hand, a particle on the equator, at rest relative to the Earth, is not rotating at all in the horizontal plane, and its vertical component of vorticity is zero. Thus (Fig. 23) the vertical component of the Earth's vorticity

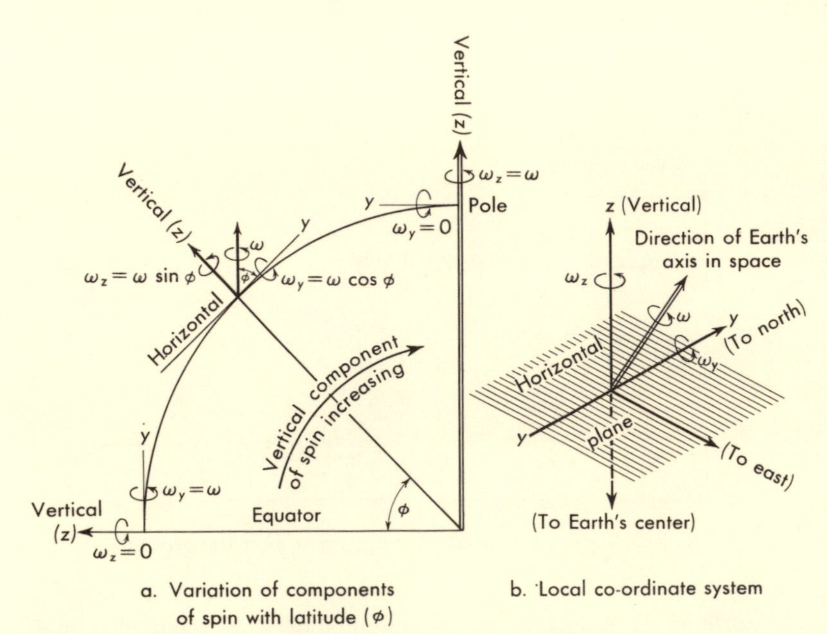

Fig. 23. How the vertical component (ω_z) of the Earth's spin (ω) changes with latitude.

depends on latitude and changes continuously from 2ω at the poles, where the latitude is 90°, to zero at the equator, where the latitude is 0°. The simplest function of latitude ϕ that behaves in this way is $2\omega \sin \phi$ and this, in fact, is the correct expression. This quantity, it should be noted, does not depend in any way upon the atmosphere. But the air moves over the surface of the Earth as well as rotating with it, and the *relative vorticity* of the wind is defined to be twice the angular velocity of a particle of air about a vertical fixed in the ground. Thus

$$\text{absolute vorticity} = \text{relative vorticity} + 2\omega \sin \phi$$

where, as usual in the dynamics of large-scale systems, the word "vorticity" means "the vertical component of vorticity." The principle of the conservation of absolute vorticity is that in the simplest type of atmospheric motion *a particle of air keeps its absolute vorticity unchanged as it moves over the Earth.* This property of atmospheric motion is of the greatest importance in dynamical meteorology.

The concept of divergence is, if anything, even simpler than that of vorticity. In Fig. 24 are shown examples of the horizontal streamlines of two types of motion, known as *convergence* and

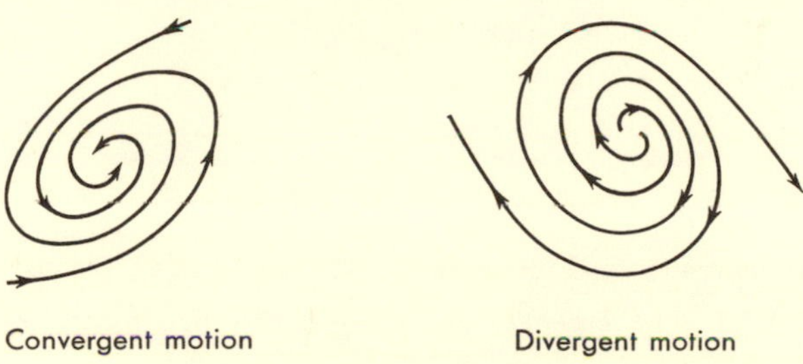

Fig. 24. Convergence and divergence.

divergence, respectively, which are especially important in meteorology. In convergent motion fluid moves into a central region, and in divergence the reverse occurs—the fluid moves away from the center. If the fluid is *incompressible,* that is, if its density is not changed by the motion, it is evident that neither pattern could exist without vertical motion, for the fluid could not accumulate in, or vacate, the central region indefinitely. Compressibility in unconfined fluids such as the atmosphere becomes significant only at speeds in the neighborhood of that of sound waves (about 700 miles an hour in air at terrestrial temperatures) or with very great accelerations, and we are unlikely to incur the risk of serious error if we treat the atmosphere as incompressible when dealing with the effects of horizontal motions of the speed of natural winds. Figure 25 shows an idealized vertical cross section of the lower levels of the atmosphere with convergent and divergent motions, together with the physical effects associated with such movements. Air flows into the lower levels of a region in which pressure is falling, but it cannot accumulate there and is forced to ascend. As it does so it cools adiabatically and condenses its water vapor into clouds, rain and snow. In a region of rising pressure the air slowly descends and flows out at the lower levels. The subsiding air is warmed by adiabatic compression, with the result that there is a general "drying out" and thinning of cloud. Deepening depressions

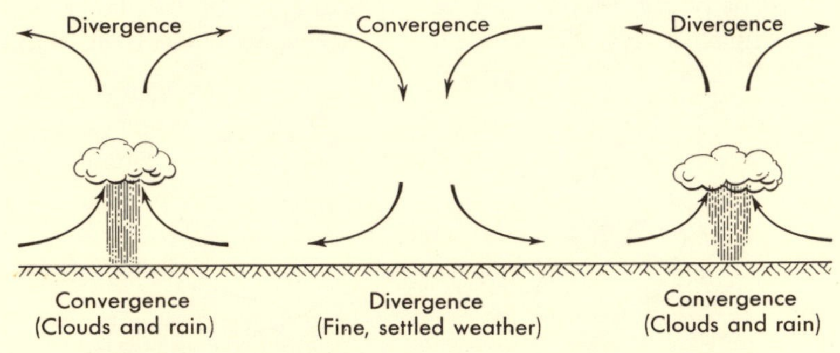

Fig. 25. Convergence and divergence in the atmosphere.

are thus associated with convergence and intensifying anticyclones with divergence, and much of the dynamical theory of weather systems turns upon this fact.

Like vorticity, divergence can be expressed mathematically in terms of the rates of change of the wind components with distance.[3] It is thus possible, in theory, to construct a map showing how divergence is distributed, but to do this satisfactorily requires accurate and plentiful wind observations. Such maps, when constructed, show regions of positive and negative divergence, and it is evident that there is a real relation between divergence and weather.

The principle of the conservation of mass is expressed by equat-

[3] See Appendix II.

ing the divergence of the velocity to the rate of change of density with time. In fluid mechanics this statement is called the *equation of continuity*. If the fluid is incompressible, the density does not change as a result of the motion, so that the divergence of velocity is zero.

It is an unfortunate fact that the mathematical term "divergence" embraces both "divergence" and "convergence" in the meteorological sense. Thus a mathematical statement that the integrated divergence of the winds in a column of air extending to the top of the atmosphere is zero does not exclude the possibility that there is convergence at some levels and divergence (in the meteorological sense) at other levels. This, in fact, must often happen. A barometer at ground level indicates the total weight of air in the column, and if the barometer is falling it is clear that the column is losing mass despite the fact that air is flowing in near the base. Thus, in the meteorological sense, there must be divergence above developing cyclones and convergence above intensifying anticyclones in order that the principle of continuity be satisfied. This principle was first enunciated by the British meteorologist W. H. Dines and is sometimes known as the *Dines compensation*.

If the vertical velocity is zero, the expression for the divergence becomes what is called, in meteorology, the *horizontal divergence* of the wind. In a strict geostrophic system the vertical velocity is always zero and the horizontal divergence of the winds vanishes. Geostrophic winds are thus nondivergent, but they may be *confluent* or *diffluent*. Convergence and divergence refer essentially to the accumulation or depletion of mass in a volume of fluid; confluence and diffluence refer to the geometrical pattern of the streamlines. When the streamlines crowd together, as at the mouthpiece of a trumpet, the motion is confluent. When they spread out, as at the exit, diffluence occurs. In a geostrophic system the speed of the wind automatically increases as the isobars become closer, just as the speed of water flowing through a pipe increases as the pipe narrows. There is a similar effect, but in the reverse direction, as the spacing between the isobars increases. Convergence and divergence in the atmosphere are indissolubly linked with the phenomena of weather through vertical motion, and the fact that a system in strict geostrophic balance is nondivergent underlines the statement made before that in such a system of winds there could be no weather.

We have now the essential tools for a discussion of the dynamics of weather systems, but before describing their use we must look at the chief features of such systems, see what are the facts that call for explanation, and learn how the models that have been developed can lead to quantitative results that are useful in the problem of prediction.

The Distribution of Weather Systems

In the hierarchy of dynamical systems we have now to consider what meteorologists have found out about those which are related more intimately to weather than the great streams which compose the general circulation. If the surface of the Earth were smooth and uniform, the distribution of pressure would be relatively simple —low near the equator and about latitude 60° and high at about latitude 30° and near the poles. How far is this scheme representative of reality? Observations show that the pattern is followed much more closely over the oceans than over the continents, chiefly because solar radiation is absorbed and stored up to a greater extent by the sea than by land, and also because of the influence of mountain ranges on the circulation.

In spring, the surface layers of the land heat up more rapidly than those of the sea, and because warm air is lighter than cold air there is a marked tendency for areas of low pressure to develop over the continents in summer. In the fall, the land cools rapidly and great cold areas of high pressure (anticyclones) form over the continents, with low pressure over the relatively warm oceans. This causes the pressure pattern and the zonal motion described in Chapter 2 to be modified, especially in the Northern Hemisphere, which contains the great land masses of Europe, North America and most of Asia. In the Southern Hemisphere, where the oceans cover a far greater part of the surface, the pressure pattern is simpler and conforms more closely to that described in Chapter 2.

In the Northern Hemisphere winter, the polar high-pressure area tends to be displaced toward the Alaskan mountains, which form an effective barrier against intrusions of warm air from the Pacific. In North America there is usually a cold anticyclone just east of the Rocky Mountains and, in general, cooling is so intense that the subpolar low-pressure belt is wiped out over land but is intensified over the seas to form the Icelandic low in the North Atlantic and the Aleutian low in the Pacific. In the Southeast the high-pressure area over North America merges into the subtropical anticyclonic belt. In summer, the polar high-pressure area is central and the Icelandic and Aleutian lows are less deep. The subtropical high-pressure belt is well in evidence over the oceans but not over the continents, where pressure is low over the heated land. In America there is a well-marked low-pressure area covering Arizona and northern Mexico.

Although the distribution of cyclones and anticyclones over the globe is by no means uniform, it is possible to recognize certain recurring features. The tracks of most cyclones lie between 50° and 60° north and south of the equator. The zonal distribution is especially well marked in the Southern Hemisphere, but in the Northern Hemisphere the regularity is upset by the great land masses which tend to breed anticyclones.

In general, cyclones move from west to east and two specially favored winter tracks are in the Pacific, from Southeast Asia to the Gulf of Alaska, and in the North Atlantic, from Labrador to the Greenland and Icelandic seas. The Pacific cyclones are held up by the mountain barriers on the west of the United States and Canada, but inland there are regions with a notably high frequency of cyclones in winter. These are chiefly east of the mountain ranges of the Sierra Nevada, the Colorado Rockies, and the Canadian Rockies, and also in the vicinity of the Great Lakes. The Colorado cyclones favor a northeasterly track toward the Great Lakes, and during their passage they dominate the weather of the central and eastern parts of the United States. The Alberta storms, originating east of the Canadian Rockies, often drag cold air over the Great Plains during their eastward passage. Cyclones are particularly frequent near the Great Lakes in winter, partly because so many tracks pass over this region and partly because of the high temperature of the water relative to the land.

In summer, the position is more complicated. There is a northern belt of cyclonic activity in the vicinity of the boundary of the polar high-pressure air, with a characteristic sequence of weather described later in this chapter. Farther south, chiefly over Nevada, southern California, Arizona and northern Mexico, heat lows are a common feature of the lower levels of the atmosphere, but there is usually high pressure aloft and the weather sequence differs from that typical of the mid-latitude depressions. During summer there is a marked tendency for hurricanes, areas of very low pressure of smaller lateral extent than the extratropical cyclone, to form on the edge of the trade-wind belt, notably in the Caribbean seas. We discuss these in detail in Chapter 5.

The most clearly marked feature of the distribution of anticyclones in winter is the subtropical belt, especially in the Northeast Pacific. The United States shows a very high frequency of winter highs over Nevada, Utah and Idaho, and in general pressure is high from Alaska to the Great Plains. The great Siberian anticyclone dominates the weather of the European-Asian land mass. In summer, the influence of the Great Lakes on American weather is again seen, for the relatively cool water produces a high frequency of anticyclones.

Attempts have been made by some writers to correlate the world distribution of pressure with the intellectual and material advancement of nations and it has been argued that predominantly anticyclonic climates, in which the weather tends to be passive, with long sustained periods of extreme heat or cold, are less conducive to mental and bodily vigor than is the climate of the cyclonic regions, in which changes occur every few days. This view of history was put forward with eloquence by the American geographer Ellsworth Huntington, and there can be little doubt that climate has played a large part in shaping the destiny of peoples, especially in the early days of civilization. But other factors, such as ease of intercourse with other races and abundance or lack of natural resources, must also be considered and the question is too complex to be resolved by an appeal to climate alone. Today technology, in the shape of air conditioning, has gone far toward remedying the deficiencies of nature.

Air Masses

One effect of land masses is to break up the major belts of high pressure into discrete anticyclones. Apart from the polar highs, the most permanent features are perhaps the Pacific and Azores anticyclones, but these are by no means stationary. They are subject to seasonal movements, toward the equator in the summer and poleward in the winter. The size, intensity and position of these anticyclones relative to the polar anticyclones exert a large influence on our weather, but to follow this process in detail it is necessary to consider what happens when such great volumes of air, of widely different physical characteristics, come into contact.

Most of the rain and snow that falls in the temperate zones of the Earth is associated with the movement of limited areas of low pressure called *cyclones, lows* or *depressions.* In 1863, when meteorology was in its infancy, Admiral FitzRoy, the first director of the Meteorological Office of Great Britain, put forward the idea, then revolutionary, that such depressions are formed by the meeting of two great uniform air streams with very different properties. One current, cold and dry, comes from the polar high-pressure area and the other, warm and humid, from the subtropical high-pressure belt. The typical depression is formed on the boundary between the two streams. In the language of modern meteorology, FitzRoy had traced the origin of the extratropical cyclone to the meeting of two *air masses* and had correctly placed the site of the storm on the *front,* or boundary between the masses. FitzRoy's ideas lay dormant for the next half century and did not receive their final form until the brilliant work of the Norwegian meteorologists during World War I made precise much that was latent in his work.

In the language of modern meteorology, an air mass is a body of air, of horizontal scale of the order of thousands of miles, within which there are no sharp systematic differences of temperature and humidity. The sources of the main air masses are the anticyclones that make up the subtropical and polar high-pressure belts. The air circulating in these systems moves slowly and has time to acquire uniformly the properties of the underlying surfaces. The simplest possible classification of air masses is thus into air of "tropical" and "polar" origin, but as an air mass moves away from its source it is modified by the surface over which it passes and a further subdivision into "continental" and "maritime" masses is therefore necessary. The life history of air masses plays an important part in the analysis of weather charts, especially in relation to stability. The effect of a warm surface on a cold air mass passing over it is to form a steep lapse rate of temperature in the lower layers, with the result that the system becomes unstable with gusty winds and a tendency to thunderstorms. A warm mass passing over a colder surface forms a low-level inversion and becomes stable, with steady winds and generally quiet weather.

In winter, the chief air masses that influence the weather of the North American continent are of continental-polar, maritime-polar and maritime-tropical origin. Continental-polar air usually moves from the north of the continent southward over the Great Lakes (where it picks up water vapor) and then southeasterly. Such movements generally bring snow on the eastern shores of the Great Lakes and in the mountains of the Appalachian system, but clear weather farther east. A less favored path takes the air west of the lakes and often brings clear cold weather to the plains. The Pacific maritime-polar air meets the barrier of the Rockies and brings cloud and precipitation to the windward slopes. The air that passes over the mountains loses much of its moisture in the ascent and is heated by adiabatic compression as it descends on the lee. This process explains the Chinook wind which the Blackfoot Indians called the "snow eater"—with good reason, for the dry warm air can evaporate a snow cover in remarkably short time. Maritime-tropical air usually comes from the Gulf of Mexico. It is warm and humid, and stable over the cold land. If it meets a cold air mass, rain occurs as the result of a process of lifting which will be described in detail later, but the air mass usually turns eastward into the Atlantic and rarely penetrates into the northerly regions in winter.

In summer, continental-polar air, Pacific and Atlantic polar air, and maritime-tropical air from the Caribbean region affect the United States. Maritime-polar air from the Pacific may bring low cloud and fog to the California coast. The hot and humid Caribbean air is unstable over the southern coastal areas, and heavy showers, thunderstorms and tornadoes may result. Polar air masses usually undergo considerable changes as they move over land, whereas tropical air changes little, but like all generalizations in meteorology, there are many exceptions to this rule.

Fronts and Cyclones

Between 1910 and 1920 a small group of mathematicians and physicists working at Bergen in Norway used the air-mass concept to change the whole direction of meteorological thinking. Their main achievement was to make precise, and convert into a workable system of analysis, ideas which were implicit in the studies of earlier meteorologists, notably those of FitzRoy and the British meteorologists Napier Shaw and R. G. K. Lempfert, but this in no way detracts from the freshness and originality of the researches which, in a remarkably short space of time, changed the appearance of synoptic charts all over the world. Previous to this the forecaster had been accustomed to emphasizing the element of continuity in his charts—isobars were drawn as smooth curves and abrupt changes in wind and temperature were looked upon as unreal. The Bergen school went out boldly for discontinuities as the significant features of the physical structure of the atmosphere in relation to weather.

We have seen that there must exist in the atmosphere large currents of air with fairly uniform physical characteristics. What happens when two such currents flow side by side? Clearly, there must be sharp changes in temperature and humidity in a very narrow zone, but it is not to be expected that this state can continue indefinitely, just as two armies facing each other from lines of trenches are not likely to remain passive forever. Sooner or later a battle is bound to develop. In the atmosphere the battle takes the form of a cyclone (low or depression), a distinct dynamic system with a well-defined life history. Some analogy like this may have been in the minds of the Bergen meteorologists when they gave the name *front*—or, more accurately, *frontal zone*—to the line of demarkation between the polar and tropical air masses. Outside neutral Norway great armies faced each other along lines of trenches which ran from the English Channel to the Pyrenees, and from time to time battles began, raged and died away.

The fronts of the meteorologist are not only zones of discontinuity in the physical properties of the air but also regions of "concentrated weather," in that the more dramatic features of the storms are to be found in their vicinity. By their aid we can produce a rational explanation of the sequence of weather observed during the passage of a depression. The existence of the depression as an individual dynamical system of horizontal scale of many hundreds of miles had been recognized in the nineteenth century.

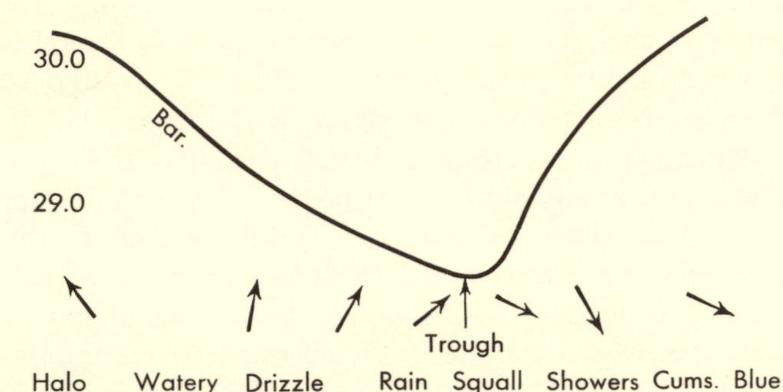

Fig. 26. Pressure and weather in a low, after Abercromby.

In 1887 the Hon. Ralph Abercromby, a gifted British amateur, published a celebrated diagram showing a depression as an oval region defined by two closed isobars, with a characteristic distribution of weather and pressure (Fig. 26). Similar schemes were used by forecasters in other countries; they were purely descriptive and essentially two-dimensional.

The theory of the Bergen school began with the concept that the genesis of a depression is a wave on the interface of two adjacent air streams of different densities. The streams were identified as part of the polar and tropical air masses. It was further assumed that the two parallel streams had different velocities—in technical language, a strong wind shear existed near the common boundary of the two air masses. In such conditions a wave forms with low pressure at its apex (Fig. 27, b). Soon the warm air develops a large bulge, called the *warm sector,* and the disturbance takes the characteristic form shown in Fig. 27, c. The surface of separation, called the *polar front,* is further divided into two segments, the *cold front* and the *warm front.*

Figure 28 is the famous diagram of J. Bjerknes and H. Solberg, published in 1921, which represents an idealized cyclone of the northern latitudes and gave the world a new conception of a depression. The lower part of the diagram shows a cross section south of the center, from which it will be seen that both fronts are gently sloping surfaces. The cold front lifts the warm air as by a wedge, and at the warm front the warm air climbs over the cold air. The upper part of the diagram shows a cross section north of the center,

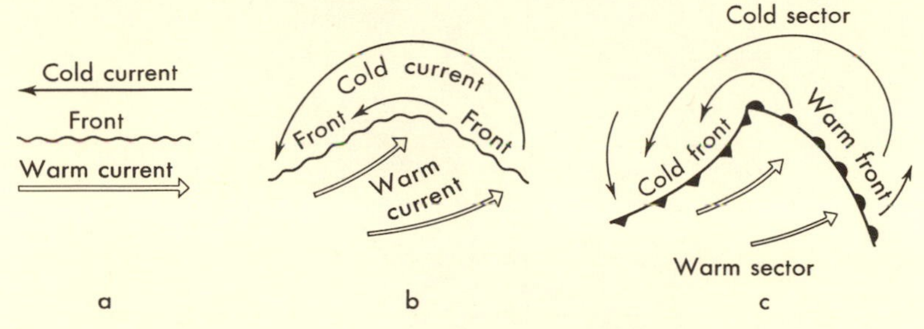

Fig. 27. The Bergen theory of the genesis and early development of a cyclonic depression.

with the warm air above the cold air. The whole system moves in a west-to-east direction, but undergoes considerable changes as it progresses.

An observer stationed to the east of such a depression experiences a well-defined sequence of weather. The first warning of the approach of the disturbance is the appearance of a belt of cirrus cloud driven by westerly winds at great heights far ahead of the storm. (This accounts for the many proverbial sayings that "mare's-tails" in the sky presage bad weather.) As the warm front approaches, the clouds thicken and become lower, the barometer falls steadily, the wind backs[4] abruptly and freshens, and the temperature rises slowly. In the frontal zone itself the clouds become low and dense, and there is continuous rain or snow some miles in advance of the warm front at the surface. With the passage of

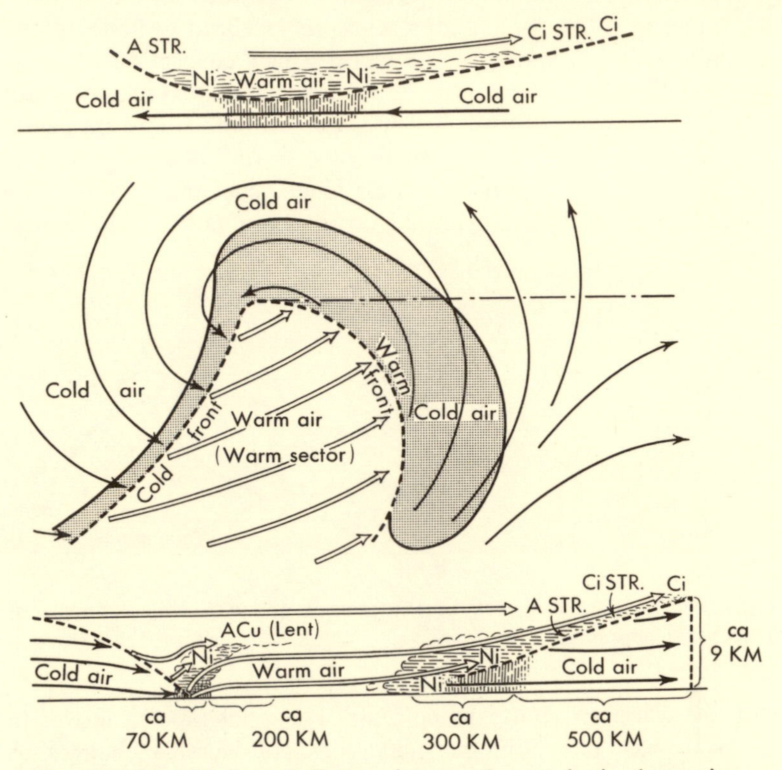

Fig. 28. The Bjerknes-Solberg picture of a cyclonic depression.

the front the fall of pressure ceases, the wind veers rapidly and slackens, and the continuous precipitation stops. In the warm sector the weather depends upon the stability of the air mass, but is generally fair with some rain or drizzle, moderate uniform temperature, and poor visibility. In the vicinity of the cold front the

[4] Changes direction in a counterclockwise fashion. A change in the opposite direction is called a veer.

nature of the weather depends to some extent upon the sharpness of the discontinuity and its speed of movement, but usually there is a quick rise of pressure accompanied by a sudden fall of temperature and change of wind direction. The front itself may be preceded by a "line squall," with a rapidly veering wind which suddenly increases to considerable strength, black menacing clouds, and torrential rain. Cold-front weather is usually of the showery type indicative of instability. With the passage of the front the clouds lift and the weather becomes fine except for occasional showers. Visibility is usually good in the cold air.

In the final stages of the life of the cyclone, which may take several days, the cold front overtakes and joins up with the warm

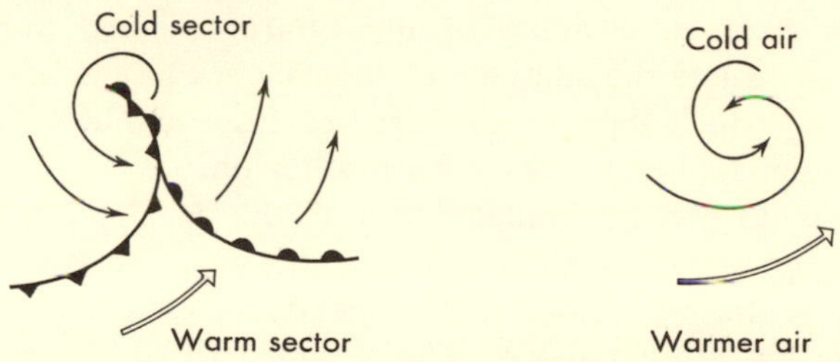

Fig. 29. The occlusion and the end of the depression.

front. The warm air is then entirely lifted from the surface and the depression is said to be *occluded*. Finally, the whole disturbance degenerates into a gently rotating mass of homogeneous air and ceases to have a distinct identity on the weather map (Fig. 29).

The sequence of weather described above has long been accepted by meteorologists as typical of the life cycle of the average cyclone or depression of the temperate latitudes. It must not be regarded as invariable and in practice it is found that relatively few depressions conform exactly to the Bergen model. The introduction of the radiosonde just before the outbreak of World War II made possible soundings even when the sky was overcast, and the wealth of information thus gained has shown that the processes involved in the formation and development of a depression are rarely as straightforward as those envisaged above. The Bergen model is an idealization, but one that has many contacts with reality, and it is by such idealizations that science progresses. Perhaps the greatest contribution of the Norwegian meteorologists was to provide a specific method of analyzing weather charts which not only helped to reduce the subjective element in forecasting but also cleared the way for the introduction of mathematical methods into this complex art.

Anticyclones

Anticyclones (highs) are regions of high pressure in which the air circulates around the point of maximum pressure in a clockwise sense in the Northern Hemisphere and anticlockwise in the Southern Hemisphere, according to Buys Ballot's law. They vary in size from the great anticyclones of the subtropical and polar high-pressure areas and the continental anticyclones to patches of high-pressure air which separate one cyclone from another. The name *ridge* is often applied to such a narrow region.

There is no specific frontal theory of the development of anticyclones comparable with that which exists for depressions. The structure of an anticyclone is less clearly defined and the whole system is far more passive. The weather in an anticyclone is usually quiet; winds are low or calm and rain is rare, but drizzle sometimes occurs. Temperature is likely to remain at about the same value for relatively long periods. In a region such as the British Isles, situated near the main track of the Atlantic depressions, a summer anticyclone comes as a welcome relief from the generally cool and unsettled weather. The high pressure, usually an offshoot of the Azores anticyclone, brings warm sunny weather with skies clear except for detached cumuli. This type of weather is responsible for the belief, encouraged by the makers of household barometers, that high pressure necessarily means fine weather of the kind that everyone welcomes. But forecasting is not as simple as this, and the barometer must be read with an eye on the origin of the air mass. Actually an astonishing variety of weather can arise from the different positions of anticyclones. For example, if the high pressure is centered over the Yukon and northern British Columbia, Canada and the central plains of the United

States experience in winter very cold weather with clear skies, but farther south the "norther" can bring blizzards. In summer, the air coming from the north is less dry and may bring thunderstorms on its southernmost boundary. A high situated over the Hudson Bay region brings maritime-polar air which is the cause of rain or snow over the eastern seaboard in winter but affords relief from the heat in the summer.

To recall some of the discussion in the earlier part of this chapter, just as a depression is characterized by ascending air, especially in the frontal zones, so an anticyclone is characterized by descending air, which brings about heating of the lower layers by adiabatic compression and the formation of an inversion, a layer of relatively warm air lying over colder surface air. This has some important consequences, especially in an industrial country. In winter, the sky in a region of high pressure is often covered with a thick layer of cloud which cuts off much of the sunshine and produces the condition known as "anticyclonic gloom." The inversion also prevents upward currents penetrating into the upper atmosphere. The winds near the center of an anticyclone are necessarily light (this is a consequence of the gradient-wind balance) and, all told, winter anticyclonic conditions are ideal for the formation of "smog" in densely populated areas. Such conditions caused the great London smog of 1952, when thousands of people died, presumably primarily as a result of breathing smoke and automobile fumes which were imprisoned in the area by the lid formed by the inversion. In a depression the natural ventilating mechanism of the atmosphere is working at its highest efficiency, but in a winter anticyclone it may cease almost completely, sometimes with dire consequences.

Dynamics of Weather Systems

The science of meteorology, if it can be described in a single phrase, is concerned with transformations of energy in the atmosphere. The cyclone is an example of the transformation of potential energy into kinetic energy.[5] In the initial stage, air masses of different densities move side by side. After the passage of the

[5] See Appendix I.

storm the warm, low-density air lies above the cold, high-density air. The system has lost potential energy, which it originally acquired from the Sun, but has gained kinetic energy in the increased force of the wind, the exchange taking place, for the most part, in the frontal zones.

One of the outstanding features of the motion of the atmosphere is its variation of type with height. In the lower levels of the troposphere the dynamical systems are usually of a closed type, such as cyclones and small anticyclones, with complicated structures. In the middle and upper troposphere the patterns become simpler and show a marked tendency to a wavelike form. These generalizations apply especially to the zone of the westerlies; in tropical regions there is evidence that the upper air contains more complex motion systems than the somewhat featureless lower stratum.

The wavelength of the upper-air patterns is very great, some thousands of kilometers, and usually there are not more than four or five major undulations around the globe. Such waves are now called long, or *Rossby,* waves, after the Swedish meteorologist who first showed mathematically that air currents can execute such huge slow oscillations simply as a result of the conservation of absolute vorticity coupled with the variation of the vertical component of the Earth's spin with latitude.

Rossby's argument is so simple that it is almost intuitive. Suppose that initially there is a uniform circumpolar current of air moving from west to east in the mid-latitudes. From time to time such a current must experience small disturbances, such as a deflection to the north. As the disturbed current moves across the lines of latitude, it preserves its absolute vorticity, despite the fact that the vertical component of the Earth's rotation, represented by the term $2\omega \sin \phi$, is increasing because the latitude ϕ is increasing. This means that the vorticity relative to the Earth must decrease and ultimately become negative, or anticyclonic. The current must then move more rapidly on its northern boundary than on its southern boundary, like a line of soldiers executing the order to wheel. It develops a curve to the right which in time brings it back to its original latitude. On crossing this latitude the reverse occurs; the relative vorticity now increases because the vertical component of the Earth's spin decreases with movement toward the equator, and the air particles develop cyclonic, or positive, vorticity. This causes a swing northward, and so on. A zonal current will thus, when disturbed, execute a regular oscillation about its original latitude because of the conservation of absolute vorticity. According to Rossby, this is how the great planetary waves of the upper air are formed.

From his study of the dynamics of waves formed in this way Rossby was able to deduce some interesting properties. First, he was able to show that the waves must move more slowly than the zonal wind speed. They may be progressive (moving with the current), retrogressive (moving against the current), or stationary. If the zonal wind speed is between 4 and 20 meters a second, the wavelength of the stationary waves lies between 3,700 and 8,300 kilometers in latitude 60°. This estimate of the wavelength agrees with the observations, and it may be accepted that the variation of relative vorticity alone is capable of affecting atmospheric phenomena on the scale of the mid-latitude cyclones and anticyclones.

Rossby's analysis also shows that in a given latitude there is a critical zonal wind speed for which a disturbance of given wavelength becomes stationary. In latitude 60°, for example, the critical wind speed for a wavelength of 4,000 kilometers (about 2,500 miles, a typical value) is about 5 meters a second (about 10 miles an hour). Waves of length shorter than the critical are progressive; longer waves are retrogressive. This criterion is useful for upper-air forecasts at about the 500 mb (18,000 feet) level.

It is possible to regard the formation of long waves as part of the energy-transfer process of the atmosphere. In recent years much attention has been paid to the manner in which irregular disturbances (eddies) form in a current of air, and it is now thought that in nearly all instances natural motion consists of a basic mean flow with a wide range of subsidiary motions with different scales of length, and that energy continually passes from the large to the small length-scales. This is the fundamental concept in the theory of turbulent flow put forward by the Russian mathematician A. N. Kolmogoroff. A large-scale motion in the atmosphere cannot persist unchanged indefinitely, for it is unstable to small disturbances and ultimately must break down into smaller motions. The limit of subdivision is reached when the energy is absorbed into the motion of the molecules by viscosity and is manifest as heat. In the same way we may consider that the bulk of the energy of the atmosphere is stored in the general circulation and that the long planetary waves represent the first step in the passage of energy from the main storehouse to the ultimate consumers, the small eddies which turn kinetic energy into heat and thus help to maintain the radiation balance with the Sun. In Kolmogoroff's theory there is a range of eddies the characteristics of which are determined by the rate at which energy passes from the great storehouses and not by the ultimate rate of dissipation by the small motions. There is perhaps a parallel here with the weather-producing systems of the atmosphere.

We come now to the weather systems proper. The cyclone is characterized by a large fall of pressure at the surface and continuous widespread precipitation. We have already discussed the mechanism of these features in broad terms (p. 82) and here we shall look at the matter in greater detail. Atmospheric pressure is a measure of the mass of air above the surface of observation, and changes of pressure are the result of horizontal mass transport over the surface together with any vertical mass transport through the surface. It is therefore understandable that there is a simple mathematical relation between the time rate of change of pressure at any level and the divergence (in the mathematical sense) of the winds. At the ground, where the vertical velocity is zero, this relation takes a particularly simple form, known as the *tendency equation,*[6] which says that the time rate of change of pressure at the surface (the "barometric tendency" in the language of meteorology) is decided by the sum of all the horizontal divergences of the wind in the column of air above the point at which the pressure is measured. Observation shows that except in very small-scale disturbances such as tornadoes the rate of change of pressure at the surface is small, not more than a few millibars an hour and usually very much less. The tendency equation then implies that the total divergence in a column of air also must be small, so that to balance a large convergence of air in the lower levels of a cyclone there must be a slightly greater divergence aloft. There may, in fact, be convergence and divergence at various levels as long as the

[6] See Appendix II.

final result is a small net loss of air or divergence (in the meteorological sense). Thus the Dines compensation and the mathematical reasoning put into precise form the intuitive arguments of p. 82 concerning the apparent paradox that air flows into the lower levels of the atmosphere in a region of falling pressure. The equation also demonstrates the restrictive nature of the principle of the geostrophic balance, for in such a system the horizontal divergence of the wind is zero, which means that the barometer would remain motionless.

The conversion of potential energy into kinetic energy in a cyclone is called *development*. The difference between the actual wind and the theoretical geostrophic wind, which is clearly of importance in the dynamics of weather systems, is called the *ageostrophic wind*. If development is to take place, the wind must not blow exactly along the isobars. There must be a component directed toward or away from the center of the system. Some deviation from flow along the isobars in the lower layers is caused by friction with the ground, but calculation shows that this effect is not large enough to account for the convergence required to produce the widespread heavy rain or snow of a vigorous cyclone. With curved isobars there is an acceleration toward the center caused by the air being constrained to move on a curved path, giving rise to the gradient wind (p. 40) as distinct from the geostrophic wind. Unless the isobars are true circles, which is unlikely, there must be some convergence or divergence with the gradient wind, but again the effect is not large enough to explain the characteristic weather of a cyclone.

There remains the effect of the changing pressure on the winds. Lines on a weather map joining points with the same rates of change of pressure are called *isallobars*, and in practice it is found that the isallobars fall into regular patterns not unlike those formed by the isobars. It is thus possible to identify on a weather map isallobaric "lows" and "highs," regions of maximum rates of fall or rise of pressure, and to measure the isallobaric gradient, just as the pressure gradient is measured. In the usual derivation of the geostrophic wind formula no allowance is made for change of pressure with time, and when this is taken into account the result is to add to the geostrophic wind a component blowing across the isallobars at right angles and directed toward the isallobaric low. The magnitude of this ageostrophic wind is proportional to the gradient of the isallobars. This means that there is convergence into an isallobaric low and divergence out of an isallobaric high, a deduction which is supported by the observation that there is usually heavy cloud in areas of rapidly falling pressure, while clear skies or only thin cloud are found to accompany rising pressure.

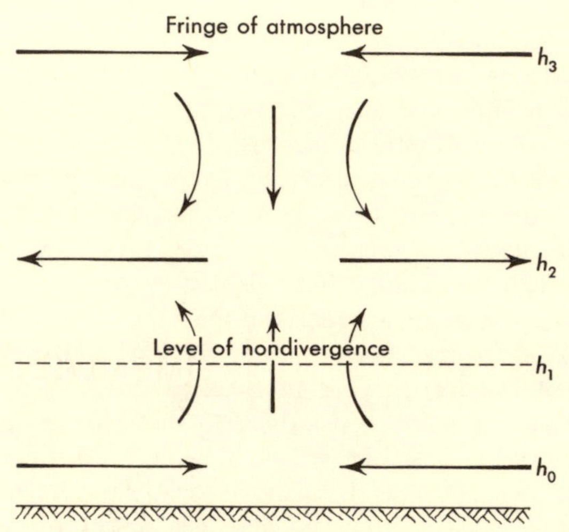

Fig. 30. Sutcliffe's development theory.

It is thus possible to account, in general terms, for the ageostrophic winds that cause convergence and hence upward motion in a cyclone. The next significant step in the dynamics of weather systems was made in 1947 by the British meteorologist R. C. Sutcliffe who, starting from the principle of the Dines compensation, gave a quantitative account of development in terms of vorticity and thermal winds.

In Fig. 30, h_0, h_1, h_2 and h_3 indicate approximately horizontal surfaces in a column of air which extends from the ground to the top of the atmosphere. The surface h_0 is supposed to be very near the ground and h_3 at a great height. If the column is part of the air in a cyclone, there is convergence on h_0 and divergence on h_2, with rising air between. At the top of the column (h_3) the air converges again. Between h_0 and h_2 there must be a level, h_1, on which the wind neither converges nor diverges and the vertical motion attains its maximum value. This level of nondivergence was implied by the arguments of p. 83. The surface h_2 is high in the troposphere and h_1

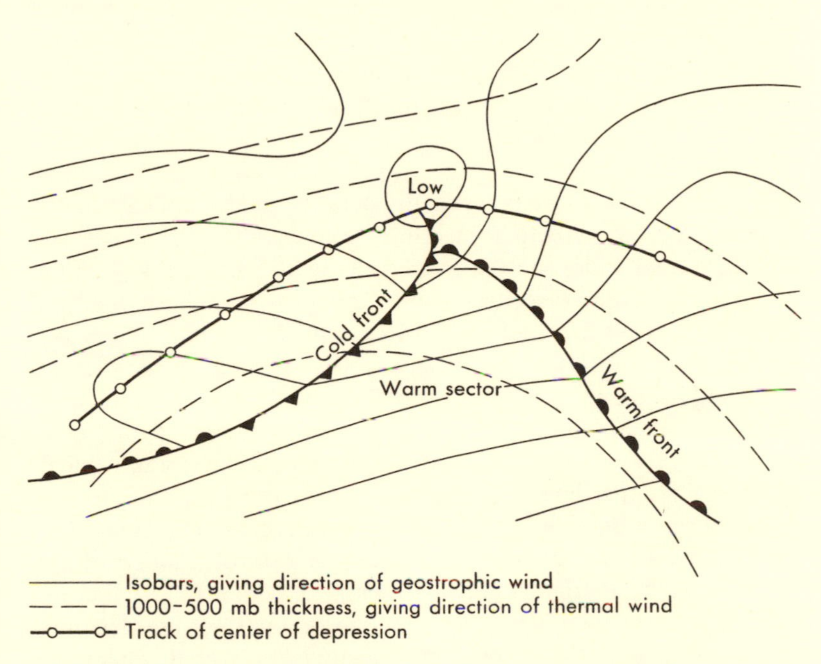

Fig. 31. Surface and thermal winds in a warm-sector depression.

may be taken to be the 500 mb surface, about 18,000 feet above sea level, or approximately halfway between the ground and the tropopause in the mid-latitudes. The variation of mass in the stratum between h_0 and h_2 must be primarily determined by what happens on the bounding surfaces, for conditions on h_3 can have relatively little effect because of the low density of the air between h_2 and h_3. Sutcliffe measures development (D) by the difference between the divergence (in the mathematical sense) of the winds on h_0 and h_2, with a sign convention to distinguish between cyclonic development (or *cyclogenesis*) and anticyclonic development. If D is positive, more air is evicted at the level h_2 than is taken in at the level h_0 and cyclogenesis occurs. If D is negative, anticyclogenesis takes place, that is, the barometer rises. The principle is therefore that of computing the difference between the low and high level ageostrophic winds as a measure of the vertical motion.

In Sutcliffe's analysis the development function D is brought into relation with the weather map by its separation into *thermal steering* and *thermal development* terms. The physical interpretation of these terms can be seen from Fig. 31 which is an idealized picture of the distribution of winds in a typical warm-sector cyclone. The thickness lines, which give the direction of the thermal wind, lie for the most part nearly parallel to the isobars in the warm sector. The "thermal steering" term involves the magnitude of the thermal wind and the rate of change of the vorticity of the surface wind in the direction of the thermal wind. The vorticity of the surface wind thus determines the sign of D, which is positive (cyclogenesis) ahead of the cyclone in the direction of the thermal wind and negative (anticyclogenesis) behind. This demonstrates the important forecasting rule that a *cyclone moves at low levels in the direction of the thermal wind.* There is, in fact, a movement of air at the upper levels through the surface-pressure pattern. The second term in Sutcliffe's dissection of the development D involves the product of the thermal wind and the rate of change of the vorticity of the thermal-wind field in the direction of the thermal wind, and thus depends entirely upon the geometry of the thickness pattern. It shows that cyclogenesis may be expected where the thickness pattern indicates a "high" of temperature (the reader will remember that thickness lines are isotherms) and anticyclogenesis near a "low." Steep gradients in the thickness pattern indicate falling pressure in warm air and rising pressure in cold air.

Sutcliffe's work is valuable in two ways: it not only offers the operational forecaster a method whereby he can identify likely areas of development on the weather map, but it also shows how the cross-isobaric flow necessary for development can be brought about without friction. As yet mathematical studies of this kind have not produced many new rules for the prediction of the behavior of weather systems, but they have made firmer the basis of many existing empirical rules and have gone a long way toward explaining why the rules work. For example, the movement of a cyclone was originally predicted from the "common-sense" rule that the direction of travel is along the line of the greatest rate of fall of pressure. With the advent of the Bergen model the rule was made more precise by the observation that a cyclone moves in the direction of the isobars in the warm sector. It often happens that the direction of the thermal wind is not markedly different from that of the isobars (in the warm sector) (see Fig. 31), but Sutcliffe's thermal-steering rule is not only more general but has a clear physical basis.

The fundamental difficulty in the application of fluid-motion theory to weather systems remains. It is that vertical mötion is virtually "unobservable" in ordinary meteorological practice. The mathematical analyses are directed to the discovery of ways of getting around this difficulty. Divergence is primarily determined by the rate of change of vertical motion with height, and in theory it is possible to evaluate the horizontal divergence directly from the weather map, and hence to compute the vertical velocity. In practice, regular wind measurements are hardly sufficiently accurate for the purpose, especially when it is remembered that considerable interpolation between widely spaced upper-air stations is called for and that the quantity sought is always small. The divergence is also approximately given by the fractional rate of change of vorticity with time (i.e., the rate of change of vorticity divided by the vorticity) and this offers a laborious but more reliable method of estimating both the divergence and the vertical motion from the weather map.

Much of the detailed analysis of the dynamics of weather-producing systems has borne fruit in the development of purely mathematical methods of forecasting pressure patterns. This aspect of meteorological science is considered in Chapter 8. We shall now pass on to systems which are smaller in scale than the cyclones and anticyclones but are associated with much greater extremes of weather.

CHAPTER 5

HURRICANES, TORNADOES AND THUNDERSTORMS

The hurricane, or typhoon, is capable of causing more widespread destruction in a short time than any other force known to man, except perhaps an earthquake and, we must now add, a thermonuclear bomb. Yet the hurricane begins its life in the tropics, in weather that is traditionally monotonous, but where the placid winds that blow over the warm seas have within them a latent force that can bring death and destruction to far-distant lands.

Like the extratropical cyclone, which we considered in the preceding chapter, the hurricane is yet another manifestation of transformation of energy in the atmosphere. The mid-latitude depression, born of the temperature contrast between two adjacent air masses, acquires its rotational energy by the lowering of its center of gravity as cold air undercuts warm air. The energy supply of the hurricane comes mainly from the ubiquitous water vapor of the atmosphere in the form of latent heat released by condensation. The life cycles of the two types of disturbance may therefore be expected to differ.

In the original scale of wind force devised for sailors by Admiral Beaufort, a hurricane was defined as a wind "such that no canvas could withstand." In modern meteorology the term is reserved for a tropical dynamical system of a special kind, characterized chiefly by the rotation of the air at speeds in excess of 65 knots (75 miles an hour) around a center of very low pressure. Such systems form over warm seas, especially in the region of the Marshall Islands and the Philippines (where they are known as *typhoons*), in the southern parts of the Indian Ocean (where they are called *cy-*

clones), in the vicinity of the West Indies, and off the west coast of Mexico. Most of the hurricanes that reach the United States form in the West Indian-Atlantic ocean. The period of highest frequency of hurricanes in this area is between August and October and, although some storms have been reported as early as May and as late as December, the true hurricane is essentially a late-summer and early-fall phenomenon. Sailors in the West Indies have a simple rhyme which expresses this fact concisely:

In June
It's too soon.
In July
Stand by!
In August
Look out you must.
In September
Remember!
In October
They're all over.

But the first and last couplets should not be taken literally.

On a weather map the fully developed hurricane is shown at sea level as a compact area of concentric circular isobars with a large gradient of pressure from the periphery to the center, where the pressure is abnormally low, generally in the region of 960 mb. (A barometric reading of 914.6 mb was reported in 1932 by a ship in the Caribbean Sea.) Ships naturally try to avoid hurricanes so that the real magnitude of the pressure drop at the center is often unknown, as is the exact shape of the isobars. It is, however, established beyond doubt that pressure at the center of a true hurricane is far below that found in mid-latitude cyclones. The winds of the system blow at very high speed, often exceeding 100 knots, around a central calm zone called the *eye of the storm* and are accompanied by torrential rain and thunder. Those who have experienced a hurricane at sea have confused memories of roaring winds, savage rain squalls and mountainous waves, like those so vividly described in Conrad's *Typhoon* and Richard Hughes's *In Hazard*. The intensity of the rain far exceeds that of a mid-latitude cyclone—a fall of 46 inches in 24 hours has been recorded in the Philippines.

Compared with mid-latitudes cyclones, the storms are of small lateral dimensions. The typical depression of the temperate zones may be between 1,000 and 2,000 miles across, but hurricanes are rarely more than a few hundred miles in diameter, and are more sharply defined. Their birthplace is the tropical ocean, between 8° and 20° latitude. During the early stages hurricanes depend on water evaporated from the warm tropical seas for their supply of energy, and if they pass over land they decrease in intensity and ultimately die away. Fortunately, relatively few hurricanes reach land. The characteristic path is initially westward on the equatorial side of the permanent high-pressure air masses and then poleward. At the westward boundary of the high-pressure system the poleward motion is usually intensified. This is the *point of recurvature* at which the track of the storm usually turns to the east, but occasionally the westward movement is maintained. (This is how some West Indian hurricanes reach the eastern seaboard of the United States.) The speed of travel of a hurricane is usually between 15 and 20 knots, which is slow enough to allow adequate warning of its approach to be given once the center has been located and the probable path diagnosed.

The symmetry of the hurricane compared with the mid-latitude cyclone enabled the early meteorologists to formulate simple rules, which have now become traditional among sailors, for the navigation of vessels in the vicinity of a tropical storm. Figure 32 shows the wind distribution in a typical hurricane. The "dangerous semicircle" is the area in which a ship may be blown toward the path of the storm. In the "navigable semicircle" the winds assist the vessel to get away from the disturbance. The first indication to the captain of the approach of a tropical cyclone is a fall in barometric pressure. With the storm several hundreds of miles away the fall is slow and may be difficult to distinguish from the normal diurnal variation[1] of pressure, but within 120 to 60 miles of the center the barometer falls rapidly and usually is unsteady. Ahead of the storm the atmosphere is unusually clear, but oppressive, with extensive cirrus cloud which does not disperse at nightfall but gives a lurid sunset. As the storm approaches, the wind freshens and changes direction and a heavy swell is noticed. A ship equipped

[1] See Appendix I.

with radar may often get confirmation of the approach of a hurricane from the characteristic echo pattern showing a spiral motion of the ring of heavy precipitation around the center.

To be reasonably safe, a ship should be at least 200 miles from the eye. To ensure this, the master must know not only the distance and bearing of the center but also the semicircle in which he is placed and the probable path of the storm. Before the days of radio this meant that the ship's position relative to the center had

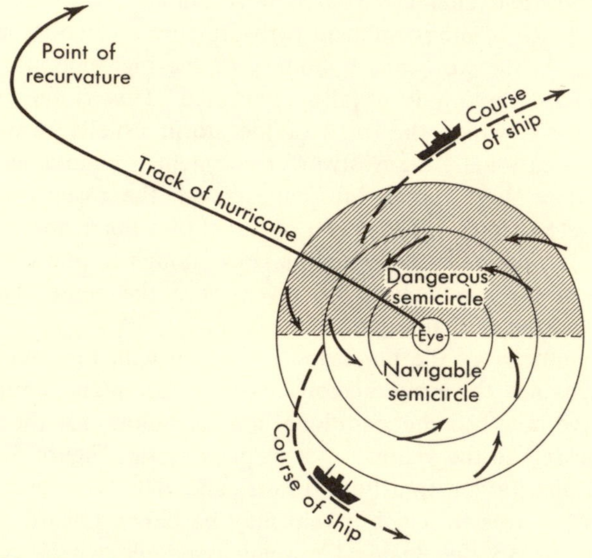

Fig. 32. Winds in a hurricane (Northern Hemisphere).

to be deduced from observations of weather in the vicinity of the vessel, and it was in these circumstances that the rules were invaluable. In the Northern Hemisphere a veering wind indicates the dangerous semicircle. The master has then with all available speed to steer a course which keeps the wind on the starboard bow. In the Southern Hemisphere a backing wind indicates the dangerous semicircle and the maneuver has to be reversed—the captain must keep the wind on his port bow. The most difficult situation arises when a ship is caught near the point of recurvature.

Today radio bulletins and reconnaissance aircraft operated by the national meteorological services have made the task of locating storms easier for those responsible for the safety of ships in tropical seas.

So much for the traditional picture of a hurricane at sea. The landsman knows the hurricane only when it crosses the coast and much of its fuel supply is cut off. Even so, the destruction wrought in settled areas can be enormous. In September, 1938, a hurricane that hit the New England area cost over 500 lives and did nearly $400 million worth of damage to property. Most of the loss of life is caused by floods and by inundation by the sea, which sweeps over low-lying coasts in huge wind-driven waves and also rises in level near the center of the storm.

We turn now to the scientific aspects. Although the hurricane is in many ways a simpler dynamical system than the mid-latitude cyclone, it has many puzzling features. In recent years American meteorologists have done much to clear up points of obscurity and a rational picture of the life cycle of these ferocious storms is now emerging. The information has come largely through the courage of those who have flown into the very heart of the storms to bring back the information needed by the scientists. There are still points of conflict in the theories, and it is necessary to select one which, although incomplete, seems to have won general approval. The following account is based largely on the writings[2] of the American meteorologist Johanne Malkus who has devoted many years to the study of tropical storms.

Hurricanes form over warm seas, usually in the western parts of the tropical oceans, where there is a deep layer of moist air near the surface. A warm sea appears to be essential, for no hurricane has ever been reported from the cooler parts of the tropical seas, such as the South Atlantic. These seas are covered by a horizontally uniform air mass swept by the trade winds from east to west. The monotony of the weather is relieved from time to time by heavy rainstorms, and the annual supply of water in the tropics is made up of relatively few downpours of high intensity. This fact suggests the existence of regular dynamical systems which

[2] In articles in the *Scientific American* and *Weather,* XIII (1958), 75, to which the reader is referred for a fuller account.

move through the tropics to disturb the traditional placidity of the weather.

In recent years American aerological studies, particularly those of Professor Riehl of Chicago, have brought to light another peculiarity of tropical weather. In the middle latitudes it is a general rule that the dynamical structure of the upper layers of the troposphere is much more regular than that of the lower layers. In the tropics the situation is reversed. There is great regularity from the surface to about 10,000 feet, but at higher levels the wind patterns, and the distribution of temperature and humidity, are often highly variable. The trade-wind current usually consists of a lower moist layer, with cumulus clouds in its upper part, and a dry cloudless upper layer. The two layers are separated by the "trade-wind inversion" which lies between 6,000 and 10,000 feet above sea level. Above the inversion clouds are rapidly eroded away by evaporation in the dry air. The trade wind grows steadily weaker with height and above 30,000 feet is replaced by a westerly current in which cyclonic and anticyclonic circulations can be identified. Such systems still form part of the troposphere, for the tropopause slopes upward from the poles to the equator and is about 50,000 feet above sea level in the tropics.

It is now well established that every hurricane begins as a mild disturbance, and one of the most difficult problems of tropical meteorology is to ascertain why so few disturbances grow into true hurricanes. The Malkus theory is that a hurricane goes through a series of crises in its history, each one of which has to be surmounted before "the storm acquires a girl's name; becomes the subject of radio broadcasts and intense aircraft reconnaissance and continues her baleful and much publicized way."

The main types of disturbances of the atmosphere over tropical seas are:

(1) *easterly waves* superimposed on the trade-wind current. On a weather map such waves are recognized by a trough of low pressure running north and south in the isobars.
(2) *tropical depressions*—centers of low pressure with winds not exceeding about 20 knots.
(3) *tropical storms*—centers of low pressure with winds above 20 knots but not exceeding 65 knots.

(4) *hurricanes*—intense low-pressure systems with winds exceeding 65 knots.

An easterly wave is shown in Fig. 33. Such waves move more slowly than the trade-wind current and bring with them a characteristic sequence of weather. Ahead of the wave, to the west, is a region of divergence with clear weather. Behind the wave, to the east, is a region of convergence with rain and cloud. In the present context the most important effect of the easterly wave is that it

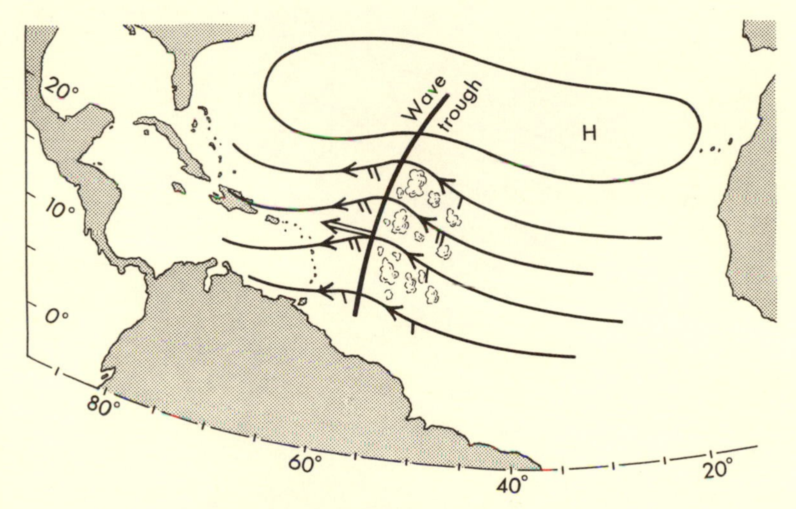

Fig. 33. An easterly wave in the tropics.

deforms and sometimes breaks up the trade-wind inversion, allowing the moist surface air an easy passage to the tropopause. This is shown by the formation of massive towering cumulus clouds in the region of disturbed weather behind the wave, instead of the normal trade-wind cumulus which does not rise above the inversion.

Some easterly waves deepen and form a closed low-pressure center or vortex (Fig. 34). This is believed to occur when the wave meets a high-level high-pressure area, which draws out air at high levels and increases convergence and cyclonic vorticity in the lower levels. The formation of a closed central area of low pressure in the wave raises the disturbance to the level of a tropical depression, or perhaps a tropical storm, but the hurricane stage has not yet been reached, and in many tropical storms the winds do not rise above the intensity of gales.

The temperature of the core indicates an important difference between the tropical storm and the true hurricane. The central region of a cyclone in the phase shown in Fig. 33 is a little colder than the surrounding air, but the eye of a hurricane is much

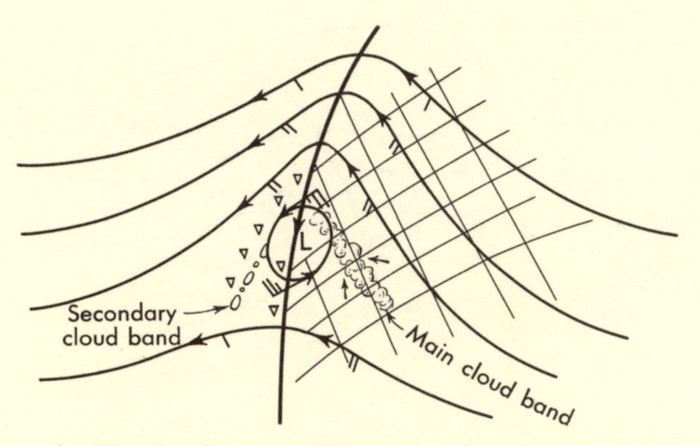

Fig. 34. Deep easterly wave with vortex.

warmer than the rest of the system. This difference appears to be critical for the build-up of the kinetic energy to hurricane level, and we must therefore look for processes by which the temperature of the central region can be raised much above that of the ambient air. Heat is released in large amounts when water vapor condenses, but a large rise in air temperature can occur only when the moist air has a chance to ascend without excessive dilution with drier air. The breaking of the trade-wind inversion provides a means whereby air can rise rapidly, as in a chimney, without too much mixing on the way up.

The rapid ascent of moist air without mixing produces temperatures in the center not more than about 5° F higher than those of the outer air currents. Apparently, this is not enough to create and sustain winds of hurricane strength. Observations indicate that the air in the eye of a hurricane is between 10° and 15° F warmer than the air circulating around it. The only mechanism capable of creating such high temperatures is the adiabatic compression of descending air.

The eye of a hurricane is a calm region with clear sky some 10 to 40 miles in diameter surrounded by a wall of dense convection clouds driven at high speed by the energy released by condensation. Outside the eye the air rises rapidly and condenses its water vapor into blinding rainstorms. Inside the eye the air subsides, and becomes warm as a result of adiabatic compression. Those who have been on board a ship which has buffeted its way to the eye speak of the atmosphere as unbearably hot and oppressive. In the lower levels the descending air is forced outward by centrifugal force to join the wild swirl around the center, so that more air is drawn in at the top to give a self-sustaining mechanism which, once established, depends for its existence chiefly upon the supply of water vapor. The formation of the warm eye is the last and most critical stage in the growth of a tropical disturbance into a true hurricane. Once the eye forms there seems to be no limit to the strength of the circulation which can develop, and it is possible that the winds attain speeds in the region of 200 knots. Over a warm water surface the system can easily obtain enough vapor to exist indefinitely, but if the fuel supply is markedly reduced, friction causes the winds to decrease and the storm to die away. The hurricane has within itself the seeds of its own destruction, for a vigorous disturbance of this kind cannot remain in the same locality indefinitely. The storm moves slowly into cooler seas or over land and is thus starved of energy. Many hurricanes, in fact, end their days as rather sedate eastern Atlantic depressions.

The chief defect in this theory is that no mechanism is proposed for the formation of the eye, and in consequence it is difficult to suggest any criterion which can decide whether or not a tropical storm is likely to become a hurricane. The necessary and sufficient conditions for the existence of a hurricane have not yet been established, but it is possible that among these may be not only a calm but also a very warm core. Whatever may be the ultimate solution, it is certain that the hurricane problem is not only one of the most important but also one of the most difficult and fascinating of the many unsolved riddles of meteorology.

Tornadoes

Although winds reach very high speeds in hurricanes, they fall far below those experienced in a tornado, which is the name given to a long narrow vortex which extends from a thundercloud to the ground. The average lateral dimension of a tornado is a couple of hundred yards and its characteristic feature is a fierce spiral motion around a nearly vertical axis. It is difficult to give precise figures for the wind speed in the vortex because no anemometer can remain unscathed in the path of the storm, but from evidence of damage it is estimated that winds of 300 miles an hour are fairly common, and it is possible that 500 miles an hour has been reached in exceptionally violent storms. Fortunately, tornadoes are usually short-lived, with a path of not more than a few miles, but during their brief lives they can be extremely destructive. A building that stands in the path of the storm is certain to be severely damaged, if not completely destroyed.

The cause of the damage is twofold: the extremely high dynamic pressure of the wind and the sudden fall in static pressure inside the vortex itself.[3] The latter effect causes houses to explode outward because of the reduced pressure outside the walls. In addition, there is a powerful upward current which sometimes causes freak effects, such as showers of frogs sucked from the ponds and deposited many miles away.

The mechanism of the tornado is still obscure, but its formation is clearly associated with marked instability in the lower layers of the atmosphere. In general, tornadoes are formed in conjunction either with strong cold fronts and squall lines or with thunderstorms. The essential initial condition appears to be a deep layer of warm moist air near the ground with a dry cold layer above. Such a distribution is very unstable, but the precise combination of conditions which produce the pendant vortex is not known. Although it is practicable to forecast the likelihood of "severe weather," it is not possible to say with confidence that a tornado

[3] See Appendix I.

will or will not form. As long as the vortex does not reach the ground the effects are not likely to be catastrophic, and there is some evidence that the "elephant's trunk" originates in a horizontal vortex in the cloud. The dark appearance of the cloud is caused by the tremendous condensation of water vapor from the sudden cooling due to expansion under the reduced pressure in the vortex. The pressure in the center of the vortex is not known with certainty, but it has been stated that a fall of 200 mb was recorded in Minnesota in 1904.

All parts of the world, except the extremely cold regions, have suffered from tornadoes at some time or another, but they are most frequent in the United States and Australia. Because of their short paths many tornadoes must go unobserved, but even so the tally is impressive. From 1915 to 1949 Kansas had 587 tornadoes and Iowa 512. In the United States nearly all tornadoes move toward the north and east, and more than half come from the southwest, with a speed of advance between 20 and 50 miles an hour. They occur at all times of the year, with a preference for late winter and spring in the southern states and for early summer in the Midwest.

A phenomenon resembling the tornado, but usually far less dangerous, is the waterspout. Waterspouts usually form over warm water, generally between May and October, but there are records of waterspouts as far north as the Grand Banks off Newfoundland in April and May. From reports of ships it seems that many waterspouts are not accompanied by high winds, but it is clear that they often exert large suction effects.

Thunderstorms

The first clear demonstration that a lightning flash is entirely similar to a long electric spark was accomplished over two hundred years ago as the result of the deliberations and experiments of Benjamin Franklin. In 1746, when he was forty years old, he constructed an "electrical machine" (now in the Franklin Museum of Philadelphia) which generated electricity by friction. With this machine and the recently invented Leyden jar (the first electric condenser), Franklin produced sparks a few inches long. He also demonstrated the lethal properties of the electrical current by killing a turkey at a picnic. (On another occasion he nearly killed himself.) These demonstrations led him to speculate whether the lightning flash, or thunderbolt, which had been the subject of superstitious dread from time immemorial, might not also be an electric spark, and in 1750 he wrote to the Royal Society of London suggesting an experiment which might settle the matter. His proposal amounted to the erection of what later came to be the "lightning rod" on the roof of a tall building in order to find out if a thundercloud passing overhead is electrified.

Franklin's suggestion was actually first carried out by a French physicist named d'Alibard, who used an iron rod 40 feet long insulated at its base by a glass bottle. On May 10, 1752, a thunderstorm broke over the little wooden hut which housed the apparatus, and an old soldier named Coiffier drew sparks from the rod by bringing an insulated wire near it. In the meantime Franklin, who had no high tower at his disposal, decided to use a kite as a means of getting information from the cloud itself. The rest of the story is familiar. Franklin drew sparks with his knuckles from a key hanging at the end of the cord that secured the kite, disregarding the risk he ran in doing so. Today we know much more about the enormous electric potentials in thunderstorms, and the reader is advised not to try to repeat the experiment. Others were less fortunate than Franklin and in 1753 a Russian scientist was killed in the course of experiments with the lightning rod. However, Franklin pursued his experiments and succeeded in showing (to use his own words) that "the clouds of a thundergust are most commonly in a negative state of electricity, but sometimes are in a positive state—the latter, I believe, is rare."

Franklin's pioneer work on thunderstorms forms a notable chapter in the history of science. In the words of Dr. B. F. J. Schonland, one of the leading authorities on thunderstorms, "[The determination of the polarity] was a beautifully planned experiment, arrived at after many years of thinking about the problem, including, no doubt, the dangers attached to the investigation, and it gave a definite answer to the question as to the sign of the charge on the base of a thundercloud. [Franklin's] statement . . . remained the only direct and reliable information on this question for 170 years. After much modern discussion and elaborate experiment it could not be put into better words today, except that for 'clouds' we would now substitute 'bases of clouds.' "[4]

The immediate practical outcome of Franklin's work was the invention of the lightning rod or conductor, which undoubtedly has saved many buildings from serious damage since it was first described, in a short paragraph sandwiched between notices of mayors' courts and announcements of Quaker meetings, in *Poor Richard's Almanack* for 1753. Franklin originally described the lightning rod as a device which attracts the lightning but disposes safely of the current. Later, in 1767, he wrote that the rod "either prevents a stroke from the cloud or, if a stroke is made, conducts it to earth with safety to the building." The view today is that the sole function of the rod is to provide a safe path of low resistance for the enormous electrical current should a stroke occur. It does not dissipate the charge in its vicinity, as was once supposed. Franklin summed up this view concisely when he wrote: "In every stroke of lightning I am of the opinion that the stream of the electrical fluid will go considerably out of a direct course for the sake of the assistance of good conductors."

By a combination of careful experimentation and shrewd reasoning Franklin thus laid the foundations of the modern theory of thunderstorms. It is appropriate that the most significant investigations into thunderstorms in our own day have been made by his fellow countrymen, using aircraft, radar, and other modern aids, with the result that we are now in a much better position to understand the mechanism of these spectacular transformations of atmospheric energy. To bring the problem into focus we must now examine the orders of magnitude of the phenomena, knowledge which was not available to Franklin when he took the first steps two hundred years ago.

A lightning stroke, or flash, is a sudden discharge of electricity between two parts of a cloud or between a cloud and the ground. The best-known form of the flash is that which occurs from the cloud to the ground, depicted for generations by artists as a brilliant zigzag line. Photographs show that the discharge actually consists of an irregularly shaped main stroke and many side

[4] *The Flight of Thunderbolts.* Oxford University Press, 1950.

branches, something like the roots of a plant or a map of a river and its tributaries. The visible length of a flash is very variable, and measurements have indicated flashes from a half to three quarters of a mile long in Europe and North America, and up to three miles in length in South Africa, where thunderstorms are frequent. Flashes of much greater length also occur within the cloud or between clouds. In contrast to its great length, the channel of the main stroke is not more than a few inches wide.

In addition to the strokes described above, there are persistent reports of "ball lightning," usually described as a moving ball of fire. It is still not absolutely certain whether such accounts refer to a real form of lightning or to an optical illusion. Undoubtedly many reports are based on a confusion between lightning and what is known as *St. Elmo's Fire,* which is an electrical discharge of the same nature as that in a gas-filled tube used for advertising. The St. Elmo glow is frequently seen on the masts and rigging of ships, and on high mountain peaks, when thunderclouds pass overhead. The prevalence of St. Elmo's Fire and the fact that, as far as the present writer knows, there has never been a convincing photograph of ball lightning, suggests that this form of electrical disturbance does not exist. This was the conclusion reached by the well-known American meteorologist W. J. Humphreys as the result of an examination of several hundred reports of ball lightning, and time has done nothing to shake his verdict.

Thunderstorms are a common feature of weather. The British meteorologist C. E. P. Brooks estimated that about 16 million thunderstorms occur over the Earth every year, or nearly 2,000 at any one time. Storms can be located at distances up to several thousand miles from an observing station by the fact that each stroke sends out a train of radio waves. Such radio "static" noises, known in meteorology as *sferics* (a shortened form of *atmospherics*), are picked up by networks of direction-finding stations operated by the national meteorological services on a routine basis. The most thundery places of the Earth are Java, Central and Southern Africa, Mexico, Panama and Brazil. The colder regions of the Earth are notably free from thunderstorms and it has been estimated that north of the Arctic Circle thunder is not heard more often than one day in ten years. On the North American continent the frequency of thunderstorms increases from north to south, reaching the maximum value in Mexico and Panama. Thunderstorms are also reported more often from the warm parts of the oceans than from the cold seas. All this points to convection as the primary cause of the storms.

The main quantitative facts about lightning can be summarized as follows: The potential differences which cause flashes to occur between the cloud and the ground and between different parts of the cloud vary between a hundred million and a billion volts. The current in the stroke varies between 20,000 and 200,000 amperes, the most frequent value being about 30,000 amperes. The time taken for the discharge to be completed varies between a hundredth and a hundred-thousandth of a second. In consequence the quantity of electricity used in a stroke is small, between 2 and 100 coulombs,[5] the most frequent value being about 20 coulombs, or twenty times the quantity of electricity which flows through a 100-watt lamp on a 100-volt circuit in 1 second. The average energy dissipated by a flash is, however, of the order of 5 billion calories. Schonland (*op. cit.*) estimates that a typical thunderstorm, producing a flash every 20 seconds, dissipates electrical energy at the average rate of 1 million kilowatts.

When a stroke occurs, the air in the narrow channels which carry the discharge is heated to about 15,000° C, or about 2½ times the surface temperature of the Sun, in a few ten-millionths of a second. An explosive expansion follows, which creates shock waves which degenerate into sound waves after traveling a relatively short distance. The sound thus produced is thunder, and observations show that a peal can be heard up to about 7 miles from the flash. The characteristic rumble of thunder arises mainly from the great length of the flash, from the multiplicity of the channels, and from echoes.

There are many mistaken beliefs about lightning, the most popular being that it never strikes twice in the same place. The records of structural damage to high buildings, especially churches, show that the belief has no foundation in fact. The lightning rod undoubtedly gives some measure of protection from severe damage, but there is no foundation for the traditional belief that it is advis-

[5] See Appendix I.

able to extinguish open domestic fires during a storm. Everyone knows, or should know, that it is dangerous to shelter under trees during a thunderstorm. The oak appears to be particularly susceptible to damage by lightning, probably because its rough bark, even when wetted by rain, does not offer a path of low resistance down which the current can pass to earth. The current must then pass through the trunk of the tree, turning the sap into steam which blasts the tree apart. Smooth-barked trees, such as the beech, are much less vulnerable, especially when wetted by rain. Even so, lightning is a serious cause of forest fires, and in the states of Oregon and Washington over 5,000 such fires were attributed to thunderstorms between 1925 and 1931.

We come now to the problem that is the main interest of the meteorologist—how thunderstorms form. It is a matter of common observation that the genesis of the thunderstorm lies in the cumulus or convection type of cloud, and that as a general rule thunderstorms form in oppressive weather, that is, in warm moist air. A thundercloud is an electrical generator and, like all such machines, works by separating the total electrical charge into equal quantities of positive and negative electricity. The simplest model of a thundercloud is that proposed by Franklin, namely, one in which there is a preponderance of negative charge in the lower parts of the cloud and of positive charge at the top. The primary problem is to discover the mechanism which causes charge separation in what is, after all, no more than a great mass of water drops and ice crystals. The answer clearly requires detailed investigations into the dynamics of the air streams within cumulus clouds.

The prime mover of the atmospheric electrical machine must be a powerful updraft, and a thundercloud may be likened to a great stack up which moist air is drawn at high speed. The actual convection process is, however, more complicated than this simple picture suggests, and it was not until after the end of World War II that the details were revealed. Nearly all the knowledge which we now possess of the dynamics of convection in thunderstorms comes from the report[6] of the Thunderstorm Project of the U.S. Weather Bureau, for which Congress made funds avail-

[6] H. R. Byers and R. R. Braham, *The Thunderstorm.* Washington D.C.: U.S. Weather Bureau, 1949.

able in 1945. Much of this work called for great courage on the part of the aircrews and meteorologists who flew into the most violent storms. As a result of their efforts we now have a tolerably clear and coherent picture of the dynamics of the average thunderstorm.

A thunderstorm is a comparatively short-lived phenomenon in which it is possible to distinguish three stages: development, maturity and decay. The main features of the development stage are shown in Fig. 35, which represents the vertical section of a typical convection cell or cumulus cloud. During this period the air within the cloud is everywhere warmer than the surrounding air, and upward currents are present throughout the entire cell. Air is brought in (*entrained,* in meteorological language) by inflow through the sides. The magnitude of the updraft increases with height and reaches high values, in excess of 30 feet a second (about 20 miles an hour) in the upper levels. In other words, it may happen that the largest component of the velocity of the air in a thunderstorm is the vertical instead of the horizontal, as in other atmospheric systems.

The initial or cumulus-building period usually lasts for about 10 or 15 minutes. During this time the cell increases its lateral dimensions from 1 or 2 miles to perhaps 6 miles and grows vertically until it reaches 25,000 or 30,000 feet. The quantity of water in the liquid and solid state increases during the period of growth, but the powerful updraft prevents much precipitation. In the dark interior of the cloud raindrops, snowflakes and ice crystals are whirled upward and downward by the turbulent convection currents, in the course of which charge separation takes place by processes which we consider later.

Figure 36 shows the vertical section of the convection cell in the mature stage. During this period the drops and ice crystals in the cloud increase in size and number as more vapor condenses in the strong updraft. In time the particles grow so large that their terminal velocity exceeds the speed of the updraft and precipitation commences, in the form of a shower of large drops and hail. The appearance of such a shower is held to mark the end of the period of growth of the cell.

From Fig. 36 it will be seen that there are now downward

currents in some parts of the cloud. These are caused by the drag of the falling drops on the surrounding air. At first the downdrafts are confined to the middle and lower parts of the cell, but gradually they extend laterally and vertically. In the center of the cloud

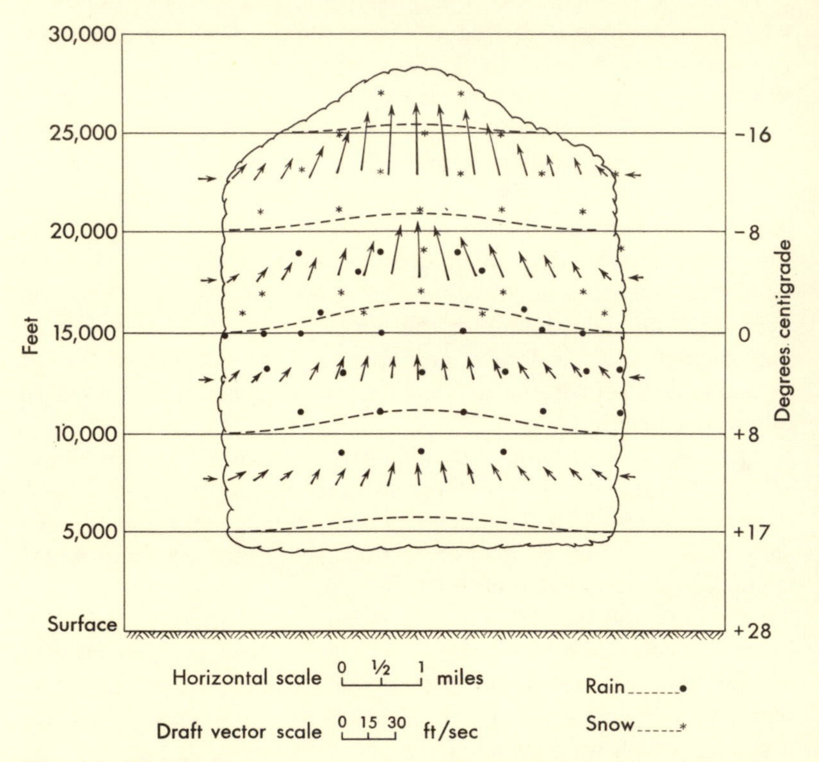

Fig. 35. The development stage of a thunderstorm (from *The Thunderstorm,* Fig. 17).

and near the top the upward motion is still very strong, and speeds up to 100 feet a second (nearly 70 miles an hour) were measured during the Thunderstorm Project. One of the most striking features of a thunderstorm is that the air moves for a short time in the vertical with the speed of a gale, making flying in the cloud not only exceedingly uncomfortable but even dangerous.

On meeting the ground the strong downdraft diverges outward, like a jet of water striking a wall, to produce a characteristic feature of thunderstorms, the gusty surface wind blowing away

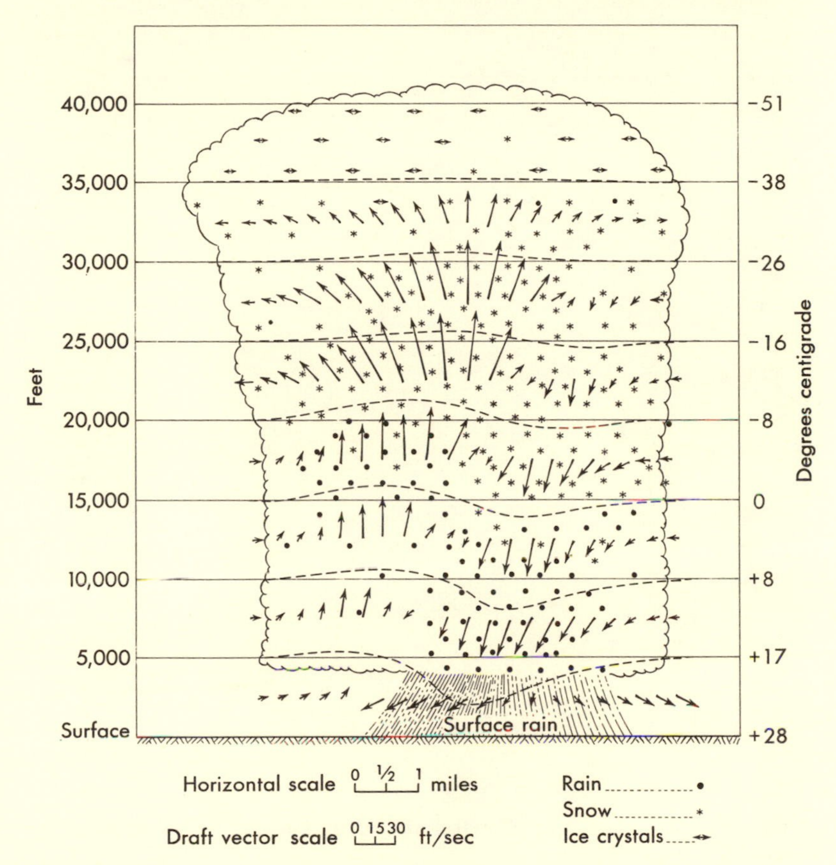

Fig. 36. The mature stage of a thunderstorm (from *The Thunderstorm,* Fig. 18).

from the area of heavy rain. At the same time the air near the ground becomes notably colder and pressure rises. Inside the cloud turbulence is often severe.

This stage in the life history of a thunderstorm lasts between

15 and 30 minutes, during which time the area of descending air increases until, at the lower levels, it finally extends over the entire cell. At the same time the clouds tower to their maximum height, between 30,000 and 40,000 feet, with sharply defined outlines.

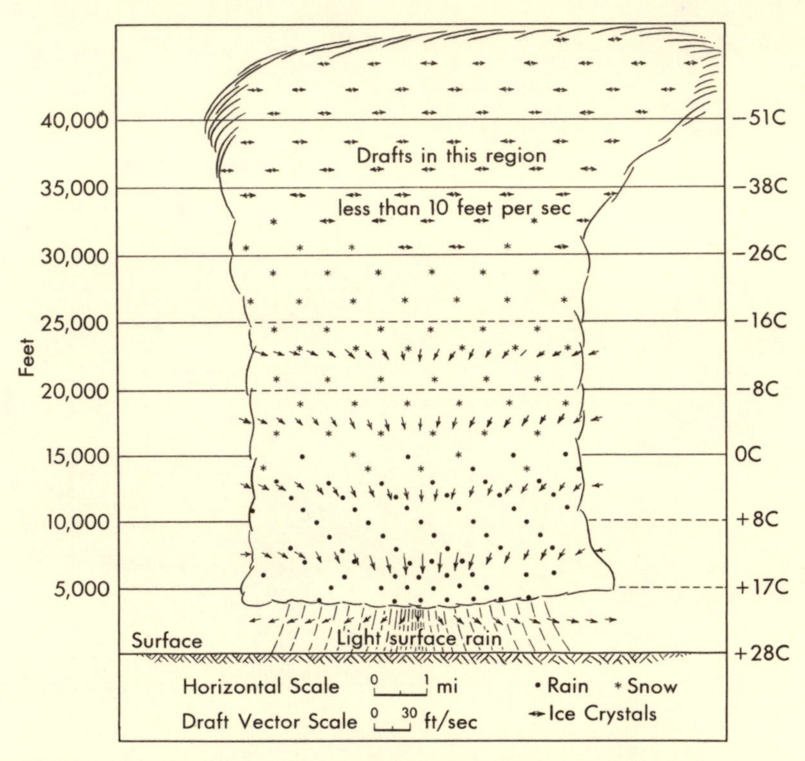

Fig. 37. The final stage of a thunderstorm (from *The Thunderstorm,* Fig. 19).

The lightning flashes are mostly from the cloud to the ground, and hail is frequently present in the precipitation.

In the final or dissipating stage the cloud becomes less sharply outlined, with wispy edges. In Fig. 37, which illustrates this phase, it is seen that the character of the vertical motion is entirely changed. The fierce upward currents of the initial and mature stages are replaced by more subdued downdrafts which ultimately spread throughout the whole cloud. As the total amount of water released by condensation is reduced, the heavy showers give way to light rain. The descending air causes a temporary fall in cloud temperature but, as the vertical motion ceases, the air within the cell gradually acquires the temperature of the environment. The weather is no longer oppressive and there is a marked feeling of relief and well-being as the last flickers of lightning, now almost entirely within the cloud, fade away. The storm generally takes about 30 minutes to die away completely.

This picture of the dynamics of a thunderstorm is, of course, idealized. Only rarely does a storm consist of a single convection cell persisting on its own. In general there are many cells in a large thunderstorm, and as each develops it rises to greater heights than those which preceded it by feeding on the saturated air of the earlier cells. Each cell has its area of upward and downward motion, and the whole system is very complex.

The energy released as latent heat of condensation is very large, and some interesting estimates have been made by the American meteorologist James E. McDonald for a small storm of 1-kilometer radius, from which it is assumed that 1 centimeter of water is precipitated.[7] The mass of water condensed is therefore at least 30,000 tons, which corresponds to the release of 2×10^{13} calories as latent heat. By an accident of figures this total is exactly that quoted for the energy release of a nominal atomic bomb of the fission type. The actual weight of water condensed is certainly greater than the amount of water that reaches the ground, probably by a factor of ten, and the latent heat of condensation does not represent the whole of the energy exchanged within the system. The calculations thus show that the energy involved in the formation of a small thunderstorm is greater than that produced by ten atomic bombs of the type used in the attacks on Japan in World War II. This assessment emphasizes the enormous scale of atmospheric processes, and is interesting in relation to fears that nuclear weapon tests can cause world-wide changes in weather.

The most uncertain part of the thunderstorm process is the mechanism of charge separation, which creates the large potential

[7] *Advances in Geophysics,* Vol. 5. New York: Academic Press, 1958.

differences that cause lightning. Here the meteorologist is for once faced with an embarrassment of riches, for there are many physical processes by which electrical charges can be separated in a mass of water drops and ice particles which is agitated by convection currents. The difficulty is to decide which process, if any, dominates. This problem has been extensively debated by meteorologists for many years, but there is still no generally accepted explanation of all the observed features of the electrification of clouds.

The starting point of all the theories is Franklin's observation that the lower part of a thundercloud has a preponderance of negative charge. This is also the region in which the larger drops are found, and Schonland suggests that a convection cell acts like a winnowing machine by separating the "chaff," the small positively charged drops and ice particles, from the "grain," the heavier negatively charged drops and hailstones. But how are the charges separated in the first instance? There are two simple ways in which a drop or ice particle can acquire an electrical charge. The atmosphere always contains *ions,* molecules which have become electrically charged by losing or gaining an electron. Atmospheric ions are formed from molecules of oxygen or nitrogen by cosmic radiation or by radium emanation from the soil in numbers sufficient to make atmospheric air slightly conducting. When a water drop is formed by condensation from vapor it has an attraction for negative ions and becomes negatively charged, leaving the surrounding air with an excess of positive ions. Secondly, if ice particles collide or rub against each other they become negatively charged by friction. The ice particles are found at the top of the cloud. As they grow larger they sink to the lower regions of the cloud, leaving the air in the upper parts with an excess of positive charge in the form of ions, small drops and ice splinters. Thus the lower regions of the cloud will contain large negatively charged drops and the upper parts positively charged small particles.

The difficulty about these two processes is that they are not likely, on their own, to create the large amounts of electricity present in a thunderstorm. The process which seems to have won most favor among meteorologists is that proposed by the Scottish physicist C. T. R. Wilson, famous for his invention of the "cloud chamber." Wilson's theory has the advantage that it suggests how the winnowing mechanism is so effective. Suppose that, as a result of the processes described previously, there are some positively charged particles in the upper regions of the cloud and some negatively charged drops or hailstones in the lower parts. An ice particle, or water drop, situated between the two regions will acquire an induced negative charge on its upper half and a positive charge on its lower half. Wilson was able to show by laboratory experiments that a negative ion in an updraft is pulled into the drop by the attraction of the positive charge on the bottom of the drop but that a positive ion is repelled by the charge on the bottom half and is swept past the drop by the air current before it is pulled in by the negative charge on the upper half. This means that an updraft reaches the top of the cloud with an excess of positive charge.

All three processes may operate at the same time and the final result is the same, a preponderance of positive charge at the top of the cloud and of negative charge at the bottom. The thundercloud is a huge battery which although momentarily discharged by the lightning flash is quickly recharged by the processes described above.

The surface of the Earth and the atmosphere have a permanent electric field. In fine weather the surface of the Earth carries a negative charge, and the upper atmosphere a positive charge. Although air is a poor conductor of electricity, it is able to pass sufficient current in the fine-weather areas to neutralize the two sets of charges in a few minutes. This does not happen, so that there must be a world-wide recharging mechanism. It was suggested by C. T. R. Wilson, and is now generally accepted, that thunderstorms play the leading part in this process. Because most thunderclouds are negatively charged in their lower regions they draw a large positive charge from the surface of the Earth, leaving it negatively electrified. To some extent this is counteracted by rain, but on balance it seems that thunderstorms and rain showers keep the surface negatively charged. The thousands of thunderclouds floating over the Earth every day act as electrical generators which maintain the electrical field despite the current flowing in the fine-weather areas. Wilson's theory also predicts that the charge on the surface should vary in sympathy with the diurnal variation of thunderstorms over the great land masses, and this has been verified by observations, especially in the clean air over the oceans and in the polar regions. One of the most puzzling features of atmospheric electricity has thus received a satisfactory explanation.

CHAPTER 6

THE MICROSCALE OF CLIMATE

In the foregoing chapters we have traced the transformations of energy in the atmosphere from the general circulation to depressions and anticyclones, hurricanes, tornadoes and thunderstorms and we now reach the last phase, the absorption of the energy into never-ending collisions of molecules. The atmosphere, as we have seen, receives most of its heat from the surface of the Earth, partly by long-wave radiation and partly by conduction. It is from the Earth also that it renews its content of water vapor, together with the nuclei of condensation and freezing, by a process fundamentally the same as conduction. This process, known generally as diffusion, is responsible for the transfer of heat, matter and momentum throughout the atmosphere and in particular from the surface to the upper layers. The study of atmospheric diffusion is, however, far from straightforward in that it involves the detailed examination of *turbulence,* the state of fluid motion which, by its complexity, constitutes the outstanding difficulty of hydrodynamics.

The term "micrometeorology" is usually taken to mean the study of the atmosphere near the surface of the Earth, the region in which life is most abundant. In practice it largely resolves itself into an examination of the spread of certain transferable entities, chiefly heat, momentum and water-vapor content, by the surface wind. The micrometeorologist is concerned in the main with such apparently diverse topics as the way in which heat is transferred from the surface to the air above by conduction and convection, the structure of the wind at low levels, evaporation and condensation, and the spread of gaseous and particulate matter such as smoke, dust and pollen. Micrometeorology deals not with the analysis and prediction of weather but with the primary physical processes on which depend, in the final analysis, not only weather but life itself.

Turbulence

Only in exceptional circumstances can fluids be said truly to move in straight lines or along smooth curves. The normal state of fluid motion is one in which the particles follow tortuous paths and the motion is never completely steady, but fluctuates from instant to instant. The difficulties of the mathematical analysis of fluid motion are so great that hydrodynamics, in its classical form, ignores these irregularities and deals only with hypothetical smooth motions which are more readily amenable to calculation. Any other motion is called *turbulent*.

Turbulence is usually described as a state of motion of a fluid in which the velocity exhibits finite fluctuations of a random character. In *laminar* motion the velocity is supposed to be free from any but infinitesimal fluctuations caused by molecular agitation. The difference between the two types of motion was first clearly demonstrated by the British mathematician and physicist Osborne Reynolds, in an experiment that has now become classical. Reynolds showed that if the motion of water in a long straight tube is made visible by the introduction of a thin stream of dye at the entrance, it is possible to establish a slow flow in which the thread of dye moves down the tube with only negligible mixing, but that when the speed of flow is increased a point is reached when the line of dyed water suddenly becomes agitated and breaks up, filling the tube with dilute color. In the initial, or laminar, state the motion is almost entirely in one direction, but in the second, or turbulent, state crosscurrents of considerable magnitude form and rapidly break up the thread of dye. In other words, turbulence is associated with an enhanced degree of diffusion attributable to the formation of persistent secondary motions. Precisely how such large random motions arise spontaneously in a steady one-dimensional stream is not yet fully understood, despite much intensive research in the wind tunnels of fluid-motion laboratories.

It is a matter of great importance in meteorology, and in other sciences, to have a means of deciding whether a motion is laminar or turbulent, other than by its appearance. From his experiments Reynolds found a simple criterion which, in broad terms, indicates that laminar motion can be maintained indefinitely only in very shallow layers or at very low speeds. Yet in the atmosphere near the ground near-laminar motion is frequently found in layers which are "deep" when compared with the diameters of the pipes used in the laboratory experiments. Clearly some factor not considered by Reynolds in his pioneer laboratory studies must be present to control turbulence in the atmosphere.

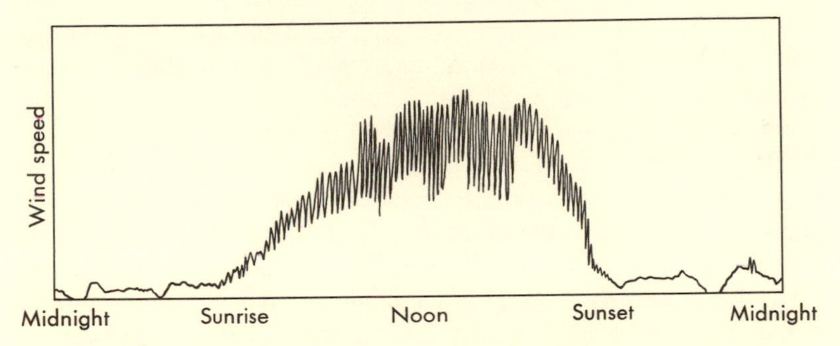

Fig. 38. Turbulent and near-laminar motion near the ground.

The counterpart of Reynolds' experiment is found in the atmosphere near the ground on most occasions of 24 hours of clear weather. Figure 38 shows the record of a sensitive quick-response anemometer sited about 30 feet above ground in open level country. Soon after sunrise the wind speed increases and rapid irregular oscillations appear in the trace. These oscillations increase in magnitude and reach their maximum amplitude in the hours around noon. In the later afternoon the oscillations become weaker and shortly after sunset they die away almost completely. During the hours of darkness the wind speed exhibits only slow changes of long period, and at times it sinks to a calm. The oscillations reappear soon after sunrise.

In terms of Reynolds' experiment, it appears that in clear

weather the surface wind is turbulent during the day and almost laminar at night. True laminar flow is never observed in the lower atmosphere. There are always small random oscillations in the wind record, caused by the roughness of the surface over which the air passes.

The existence of a marked diurnal variation in the intensity of the turbulence immediately suggests that the state of the motion must be related to the temperature of the lower layers, which also exhibits a well-marked diurnal change. Further reflection indicates that the primary cause of the variation is not the absolute temperature of the air but its mode of distribution with height. On a clear day the Sun heats the ground strongly, causing convection currents in the atmosphere adjacent to the surface. In the daytime warm air lies below cold air and any small disturbance (such as that caused by the air stream meeting a small obstacle) which pushes up the air momentarily is enhanced by buoyancy. One effect of the heated surface is thus to promote a continuous exchange of air between the lower and upper strata. On clear nights the ground cools rapidly because of the loss of heat by outgoing radiation. Air near the ground is chilled and becomes denser than the air above, and convection currents no longer exist. Any vertical motion arising from a small disturbance is soon damped out by the unfavorable density distribution, so that turbulence is suppressed and the motion approaches the laminar state. The exchange of air between the upper and lower layers is almost entirely eliminated and the particles tend to move in horizontal planes at low speed.

This picture of wind structure suggests that the vertical gradient of temperature near the ground should be closely related to the turbulence of the wind in the same layer. Figure 39 shows the record of the difference in the temperature of the air measured over a small height interval near a level surface. From just after sunrise to just before sunset, temperature falls with height, the maximum gradient, or lapse rate, being reached at local noon. During the hours of darkness the reverse condition holds—temperature increases with height, a condition known as the nocturnal inversion. This alternation between lapse and inversion corresponds exactly with the diurnal variation of the turbulence of the wind, which suggests that the basis of the theory of atmospheric turbulence should be the hypothesis that *the gradient of temperature in the vertical controls the turbulence of the wind in the lowest layers of the atmosphere*. The hypothesis is reinforced by the observation that with a completely overcast sky and a moderate or strong wind the intensity of the turbulence shows little change with time of day, and that in the same conditions the temperature gradient near the ground is also extremely steady with a value that is close to the dry-adiabatic lapse rate. (Since the dry-adiabatic

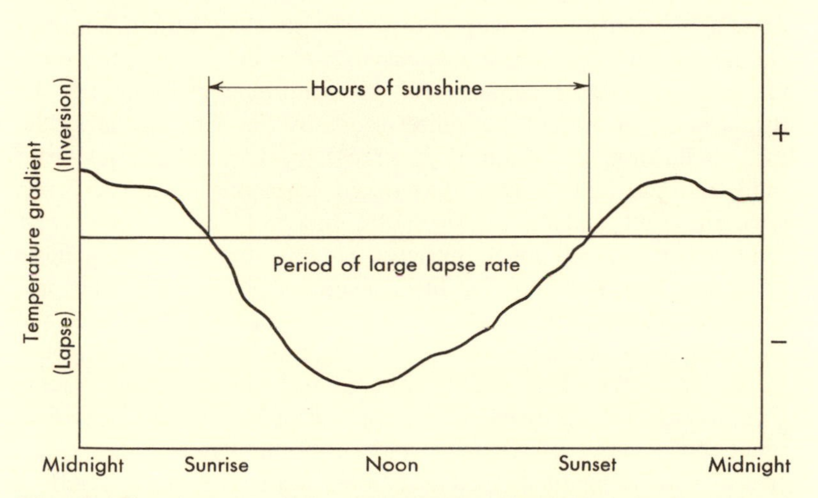

Fig. 39. The diurnal variation of temperature gradient near the ground in clear weather.

rate represents a decrease of only 1° C in 320 feet, the difference in air temperature over an interval of, say, 1 foot to 20 feet is nearly zero when the sky is completely covered with a thick sheet of cloud.)

These considerations lead to a plausible explanation of a feature of wind structure that would be little more than a meteorological curiosity were it not for the connection between turbulence and diffusion. How intimately the rate of dispersion of small airborne particles depends on the degree of turbulence of the wind is easily observed by watching the behavior of smoke from a weed fire in open country under clear skies. When the Sun is high in the sky the smoke is scattered by the wind over a wide arc, but after sunset the same fire produces an unbroken plume or sheet of smoke which may drift for long distances with little mixing with the ambient air. Diffusion proceeds at a high rate when there is a lapse of temperature and slowly in the presence of an inversion, and this has profound repercussions on the meteorology of the lower atmosphere.

The diffusion of matter by the turbulence of the wind is of the greatest importance for the maintenance of life. The cycle of water from the seas, lakes and rivers to the atmosphere, and back again as precipitation, begins with the process of evaporation, which is simply the diffusion of water vapor into the air. The Earth's vegetation cover largely depends for its continued existence on the dispersal of pollen and the lighter seeds by the wind. In an industrial civilization the removal of smoke and noxious gases from industrial and other densely populated areas is a matter of prime concern to all, and the urge to solve the problem of atmospheric pollution has given much impetus to the study of atmospheric diffusion. If it were not for the scavenging action of turbulence, the great conurbations could not exist, for man would be poisoned by his own by-products, and one of the most significant contributions that meteorology could make to the welfare of the human race would be to provide the physical basis for precise appraisement of the effects of releasing undesirable matter into the atmosphere. The siting of a nuclear reactor, for example, demands a prior calculation of the concentration of radioactive matter in the vicinity of the plant, which in turn requires of the meteorologist a quantitative theory of the diffusing action of turbulence in terms of measurable meteorological entities. We shall now see how such a theory has been constructed.

Molecules and Eddies

The development of the kinetic theory of matter by Clerk Maxwell, Boltzmann and others was not only one of the greatest triumphs of nineteenth-century physics but also the genesis of a new mode of thought in natural philosophy. The kinetic theory brought in the concept of the dynamics of large assemblies of freely moving particles, in contrast to the mechanics of systems for which the configuration could be specified. In the kinetic theory physical laws and relations are expressed solely in terms of statistical functions, such as averages.

The basic concept of the kinetic theory is that in a gas, such as air, the molecules are in incessant random motion. The average energy of this motion defines the temperature of the gas. When two samples of gas at different temperatures are mixed, the collisions between molecules lead to a sharing of molecular energy which in time becomes uniform, on the average, throughout the mixture. This is the kinetic theory explanation of the phenomenon of conduction of heat. Again, if two different gases are brought into the same vessel, molecular agitation causes the molecules to intermingle and in time there is complete mixing. This is the process of diffusion of matter. Finally, if there are fast- and slow-moving adjacent layers in a stream of gas, the random motion causes some molecules to move from the fast-moving layers into the slow-moving layers, and vice versa. In this way there is a transfer of momentum across the lines of flow, with the result that the fast-moving layers are retarded and the slow-moving layers accelerated, so that in time the frictional effect of, say, a rigid boundary spreads throughout the whole stream. The diffusion of momentum gives rise to a characteristic measurable property of a fluid called its internal friction, or *viscosity*.

The methods of the kinetic theory of matter enable us to study the consequences of the diffusion of energy (heat), matter and momentum and to deduce precise verifiable results without specifying either the configuration or the motion of the molecules in detail. In the simplest form of the theory a molecule is supposed to be an elastic particle which has mass and random velocity c (and therefore energy proportional to c^2) and travels a distance l, called its free path, before it meets another molecule. The "collision" is supposed to be like that between elastic spheres. There could hardly be anything simpler than this "billiard ball" model of the structure of a gas, yet it is capable of yielding results of considerable importance with the use of only the most straightforward mathematics.

We can illustrate the power of the kinetic theory method by considering laminar flow over a smooth plane surface. This type of motion is significant in micrometeorology, not only because it resembles the movement of the air in the lower layers of the atmosphere during the period of the nocturnal inversion but also for the more significant fact that it illustrates ideas of the kinetic theory which subsequently penetrated into the study of atmospheric turbulence with profound consequences. When a fluid moves

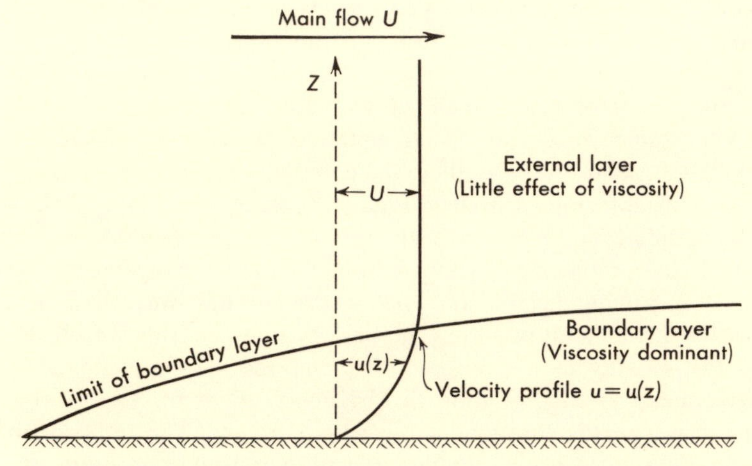

Fig. 40. Viscous laminar flow over a smooth plane surface.

over a solid surface, molecular attraction prevents any relative motion at the surface itself—even a gale is brought to rest on the surface of the ground. In the circumstances illustrated in Fig. 40 air moves over a plane smooth surface with a unidirectional velocity u which increases steadily with height. The graph of the function $u = u(z)$ is called the *velocity profile* and the rate of change of velocity with height (or du/dz in the notation of the calculus) is called the *velocity gradient*. These terms will occur frequently throughout this chapter.

Our aim is to analyze this motion mathematically. The surface exerts a *drag* or frictional force on the fluid, which must be the direct result of the internal friction or viscosity of the air. The fact that the velocity of the stream becomes progressively slower as the surface is approached shows that there must be a variable force, called the *shearing stress,* acting parallel to the direction of flow. Observations show that this force depends only upon the nature of the gas and the velocity gradient, and we are thus led to introduce into the equation defining the shearing stress per unit area a constant of proportionality, μ, called the *dynamic viscosity* of the gas. (In calculation it is usually more convenient to use instead of μ the quotient of the dynamic viscosity and the density, called the *kinematic viscosity* and invariably denoted by the Greek letter ν.)

The correctness of this approach to the problem of friction in fluids can be tested by comparison of the values of μ or ν deduced from measurements made with different experimental arrangements. If precautions are taken to avoid turbulence, the values obtained are always the same irrespective of the circumstances or the magnitude of the motion. We may then assert that the kinematic coefficient of the internal friction of air, ν, is a true physical constant with the value 0.15 square centimeters per second at 20° C and 1000 mb pressure.

This is the Newtonian or "bulk-property" approach, which accepts viscosity as a characteristic property of all fluids, prescribes how it is measured, but sheds no light on its cause. The kinetic theory approach, originated by Clerk Maxwell in 1860, is very different. It begins with the picture of a fluid composed of vast numbers of molecules darting about in a random fashion and at the same time drifting with the mean stream velocity appropriate to their position relative to the boundary. As a result of this agitation (the "heat motion"), molecules from the slower moving layers penetrate into the rapidly moving layers, and vice versa. It is supposed that, on the average, a molecule moves the distance *l,* the mean free path, without meeting another molecule, but that ultimately the slow-moving molecules collide with and retard the fast-moving molecules. Equally, the fast-moving molecules hasten the general drift of the molecules in the slow layers. In this way, molecular agitation brings about a transfer of momentum between adjacent layers of the fluid, and the rate of change of momentum, by Newton's second law, must be proportional to the force acting on the system, which is none other than the shearing stress. It is then a matter of simple mathematics to show that the constant of proportionality in the equation for the shearing stress, the coefficient of viscosity, is proportional to the product of the molecular speed and the free path, that is, to cl. The same type of argument can be applied to find expressions for the coefficients of heat conduction and diffusion in terms of the molecular constants. In every case the product cl enters but with different multipliers.

This kind of mathematical analysis of the consequences of the "billiard ball" model of a gas enables the mathematician, by making use of the observations of the physicist on the bulk properties of gases, to assign precise numerical values to the molecular constants c and l, which initially existed only as symbols in his work. Although the simple "billiard ball" model has long been superseded by more sophisticated concepts, the scientist's faith in the objective reality of molecular transfer processes is largely based on the fact that all such comparisons, although involving widely different kinds of experiments, lead to substantially the same numerical values for the molecular constants. We are thus entitled to speak with confidence of air as a gas in which, in standard conditions of temperature and pressure, the molecules move at an average speed of nearly 500 meters a second, travel a hundred-thousandth of a centimeter before meeting another molecule, and make about 5 billion such collisions a second. These figures have definite significance, although we no longer believe that molecules are exactly like tiny billiard balls, and the criterion of their reality is that they provide a satisfactory basis for the calculation of the behavior of gases in bulk.

This example of the kinetic theory treatment of viscosity has been described in detail because it illustrates the principles which led to the first breakthrough in the development of the theory of atmospheric diffusion. There are many superficial points of resemblance between the diffusing action of turbulence and that of molecular motion. If molecular forces alone were operative, mixing would be a very slow process, the reason being that, although molecules dart about at high speed, they travel only short distances before meeting other molecules and are thus very spatially limited in their action. An inspection of smoke diffusing near the ground in a turbulent wind immediately gives the impression that the scattering of the particles is brought about primarily by the relative motion of large individual masses of air, which drift with the wind but otherwise seem to move at random. An examination of the record of wind speed near the ground (Fig. 38) lends support to the idea that the oscillations are caused by the passage of innumerable whirls and vortices which, like molecules, appear to have distinct identities.

These observations suggest that the transition from laminar to turbulent flow is characterized by the rapid breakdown of the smooth motion into a mass of individual motion-systems, which are usually called *eddies,* and that the effects observed are to be attributed to the eddies, just as diffusion in a laminar stream is attributed to the motion of molecules. Obviously it is out of the question to specify either the individual motion or the configuration of the eddies, and any treatment of the problem must be statistical, following lines already established in the kinetic theory. An eddy can be envisaged as a volume of air, large compared with a molecule but otherwise of unspecified size, which for some reason or other leaves its original place in the stream and moves to, say, another level. As it moves it carries with it a content of mass, heat or momentum typical of the original layer. On reaching a new level it may be supposed to mix with the surrounding air, and in this way a much-enhanced diffusion process becomes possible. If this theory has a substantial basis in fact it should be possible to establish statistical laws for the behavior of eddies in a fluid, analogous to those established by the kinetic theory for the effects of molecular motion. In particular, in the early days of this research there seemed every possibility that the atmosphere possesses an eddy heat conductivity, an eddy diffusivity, and an eddy viscosity which, although obviously greater than their molecular counterparts, might be truly constant, that is, independent of the circumstances and scale of the process under consideration.

This view of atmospheric turbulence was much in evidence in the years between 1915 and 1935, and is especially associated with the names of the British mathematician G. I. Taylor (now Sir Geoffrey Taylor) and the Austrian meteorologist Wilhelm Schmidt. The latter published in 1925 a classic of micrometeorology with the title *Der Massenaustausch in freier Luft* (*Mass*

Exchange in the Atmosphere), which is still a valuable source of information on the effects of atmospheric turbulence.

The initial steps toward solving the problems of atmospheric diffusion were, not unnaturally, analogous with the "bulk-property" approach to similar problems in laboratory physics. It was *assumed* that all diffusion processes in the turbulent atmosphere could be described by coefficients entirely similar to the familiar molecular coefficients, that is, by constants. Schmidt used the ponderous German word *Austauschkoeffizienten* (exchange coefficients) to cover all transfer processes, and Taylor used the terms "eddy viscosity" and "eddy conductivity" for the macroscopic counterparts of the kinematic viscosity and the thermometric conductivity. The first step clearly was to establish the order of magnitude of these coefficients and the second, and more significant, step was to ascertain if the eddy diffusion coefficients were truly constant, like their molecular counterparts.

The molecular diffusivities of air (the thermometric conductivity, the kinematic viscosity, and the mass diffusivity) are all of the same order of magnitude, namely, 10^{-1} (or one-tenth) square centimeters per second. To deduce the corresponding orders of magnitude for the eddy diffusivities necessitates the use of meteorological measurements of wind and temperature and of concentration of matter released from sources of known strength. Such measurements are naturally far less precise and accurate than laboratory determinations, but they may be relied upon to fix orders of magnitude (powers of ten), which is all that is required to explore the theory. One of the best-known investigations of the early days was Taylor's determination of the coefficient of eddy viscosity, K, from pilot-balloon measurements of wind structure made over Salisbury Plain, England. We have already explained that because of friction the surface wind does not blow along the isobars but slightly across, at an angle of about 25° over land. With increasing height the angle grows gradually less until at last the wind blows more or less accurately along the isobars with the speed of the gradient wind. The height H at which the wind first attains the geostrophic direction (along the isobars) can be found with tolerable precision on any particular occasion. Over Salisbury Plain this height had been found from pilot-balloon ascents to be between 600 and 900 meters. A simple calculation shows that if molecular friction alone operated, H would be less than a meter. This proves that molecular viscosity alone cannot possibly account for the observed deviation of the surface wind from the geostrophic direction. Taylor assumed that the eddy frictional term in the equation of motion was exactly the same as the molecular friction term except that the eddy viscosity, K, replaced the kinematic viscosity, ν, and went on to calculate the order of magnitude of K which would give the observed value of H. He found that K had to be at least 10^4 (10,000) square centimeters per second, or a hundred thousand times greater than the molecular coefficient. This means that eddy friction completely dominates the approach of the surface wind of the geostrophic wind, and that molecular friction may be disregarded in studies of turbulent motion over the surface of the Earth.

Earlier Taylor had found similar but somewhat lower values for the eddy conductivity in an investigation into the cause of fogs over the Grand Banks off Newfoundland, which followed the loss of the liner *Titanic*. In this work he traced the upward growth of a layer of warm air passing over cold water. By assuming that the eddy conductivity, like the eddy viscosity, is constant and by measuring the height at which the inversion ceased, coupled with an estimate of the time during which the warm air had been flowing over the cold sea, he deduced values of the eddy conductivity of the order of 10^3 (1,000) square centimeters per second, or about ten thousand times greater than the molecular conductivity. In Austria Schmidt found similar values from a wide variety of phenomena.

These results are in harmony with the fundamental concept of the "kinetic theory" approach, for basically the same turbulence mechanism is responsible for the transfer of heat as for momentum. The somewhat lower values found by Taylor for heat transfer over the ocean were not incompatible with the values deduced for the friction effect over the much rougher land, and at this stage it seemed that Schmidt's concept of a universal exchange coefficient for the atmosphere was likely to be well founded. But doubts concerning the validity of the theory were beginning to appear, especially in relation to the upward transfer of heat by turbulence.

During a spell of clear weather the temperature of the surface undergoes a regular diurnal oscillation which, if regarded as a "wave" of temperature, can be represented as a simple sine function of time, with tolerable accuracy. The classical theory of heat conduction in solids tells us that such a sinusoidal oscillation of temperature at the surface of a body penetrates in time into the body, undergoing changes as it moves. The amplitude, or maximum value, of the oscillation decreases with distance from the surface and the time of maximum temperature becomes later. (The last-named effect is known as a *change of phase* of the wave.) The behavior of waves of temperature passing through a solid body, which has been long established in theory and confirmed in laboratory practice, had already found application in geophysics in investigations into the penetration of the diurnal and annual waves of temperature into the soil. If the eddy heat conductivity were truly constant, that is, if the air behaved as if it were a solid body as regards conduction of heat by turbulence, the diurnal variation of temperature at any height above the surface could be easily calculated from the classical formula of the conduction of heat in solids or, conversely, observations of temperature made at different heights above the surface, especially those of amplitude and change of phase of the diurnal wave, could be used to calculate the coefficient of eddy conductivity.

The surface of the Earth reaches its maximum temperature when the flow of heat into the soil exactly balances the flow of heat, by radiation and conduction, outward. The time of maximum temperature at the surface thus depends on many factors and it is not surprising that the surface of the ground attains its highest temperature some considerable time after local noon, and not when solar radiation is at its peak. It is not easy to measure the true temperature of the surface, especially when it is covered by vegetation, but the most reliable observations show clearly that the maximum is reached between 1 and 1½ hours after local noon. At the normal height of meteorological observations, about 1.2 meters above the ground, the air reaches its highest temperature much later, about 2½ hours after noon, and at heights of the order of 100 meters the maximum is not reached until some 3 or 4 hours afterward, between 5 and 6 o'clock in the afternoon. (See

Fig. 41.) The amplitude of the diurnal temperature wave also decreases with height, but not rapidly in the first 100 meters of the atmosphere. This is in marked contrast to the penetration of heat into the soil, for the diurnal temperature wave is barely detectable at 1 meter below the surface. The annual wave naturally goes much deeper and in practice can be detected at least 15 meters below the surface.

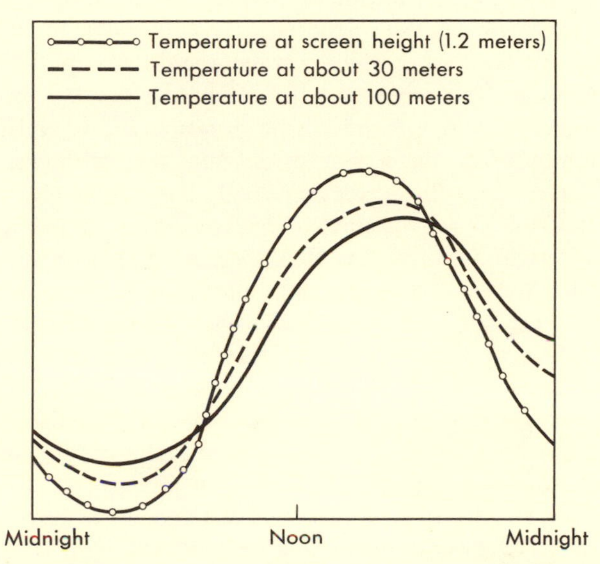

Fig. 41. Variation of the diurnal temperature wave with height (schematic).

Taylor applied this type of analysis to observations made on the Eiffel Tower in Paris, where temperature is recorded at heights of 2, 123, 197 and 302 meters above the ground. The results were somewhat disturbing in that the values of the eddy conductivity showed a marked tendency to increase with height during the summer months and to decrease with height in the winter, but it was not certain how far the Eiffel Tower thermometers were affected by radiation. Many years later the British meteorologist A. C. Best put the matter beyond doubt by showing that when very accurate measurements of the diurnal variation of air temperature made under clear skies are analyzed in this way the values of *K* derived from different height intervals do not agree. They show a rapid systematic increase with height, changing from the order of unity in the layer 2.5 to 30 centimeters above the surface to the order of 10^3 to 10^4 (1,000 to 10,000) square centimeters per second in the layer 7 to 17 meters above ground.

"Constant eddy diffusivity" finally became one of the myths of meteorology when accurate measurements of the diffusion of matter were made for defense purposes at Porton, England. When gas was first used as a weapon of war between 1916 and 1918, the opposing armies had no means of estimating the concentrations (mass of gas in unit volume of the atmosphere) at various distances downwind of the source in different meteorological conditions. They realized that inversion conditions meant high concentrations and that it was useless to release clouds of gas in strong winds and bright sunshine, but little else. At that time the meteorology of the lower atmosphere was almost as obscure as that of the stratosphere. In 1921 the British government set up an experimental station at Porton, on Salisbury Plain, to study chemical warfare, with a research group to investigate the meteorology of the lower atmosphere in relation to the diffusion of airborne substances. This group may be said to have laid the foundations of micrometeorology as an exact science, but much of the work was withheld from publication for reasons of national security, and the data were not released until after World War II.

Sources of foreign matter in the atmosphere can be classed as *instantaneous,* as in a shell burst, or *continuous,* as in the release of gas from cylinders or of smoke from generators and chimneys. Continuous emissions may be from a point, a line or an area. When the cloud from a surface continuous point source drifts downwind it spreads out in a cone, with the highest concentrations in the center of the cloud at ground level. If samples of the atmosphere are taken across the cloud at fixed distances downwind of the source, the concentration of gas or smoke can be measured with great accuracy by chemical or physical means. The crosswind distribution of concentration so obtained is shown in an idealized form in Fig. 42. From a succession of such curves, constructed from the results of trials carried out in steady conditions of small temperature gradient (overcast sky and moderate or high wind) it is possible to examine how the central or peak concentration diminishes and the width of the cloud increases with distance from the source. Additional trials were made with continuous long-line sources (across wind) to ascertain the distribution of concentration with height above ground and the fall of concentration with distance downwind, in this arrangement also.

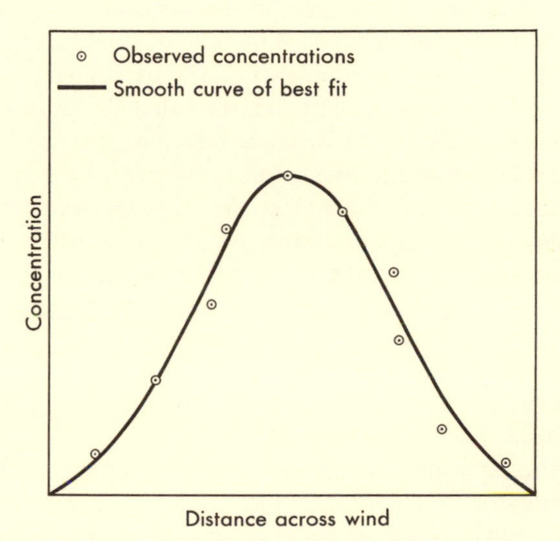

Fig. 42. Cross-wind distribution of concentration from a continuous point source.

One member of the Porton group, O. F. T. Roberts, solved the equations of diffusion for instantaneous and continuous sources of matter when the coefficient of diffusivity, K, is assumed to be constant. With these expressions it was possible to make a searching test of the "constant diffusivity" hypothesis by comparison with the results of a long series of trials made in conditions which were as comparable as is possible in the atmosphere. The results were decisive. Although the values of K required to account for the observed spread of the smoke were of the same order of magnitude as those found by Taylor and Schmidt in their studies of natural atmospheric phenomena, such as the approach to the geostrophic wind and the variation of the diurnal temperature wave with height, it was found that different values of the eddy diffusivity were needed at different distances to make the theoretical expressions agree with the observations. The reason for the apparent increase of *K* with distance from the source (which is what the comparisons revealed) soon became evident. Roberts' formulas predicted that the peak concentration in a continuous point-source cloud should decrease inversely as the distance from the source and that the width (or height) of the cloud should increase as the square root of the distance downwind. The experimental data indicated rather different laws. They showed that the peak concentration fell off inversely as the distance raised to the power of 1.76, far more rapidly than the inverse-distance law of the theory. The width of the cloud also did not conform to the law predicted by the constant diffusivity hypothesis, for the observed increase was approximately as the seven-eighths power of the distance and not as the square root. This meant that the clouds grew and were diluted far more rapidly than the formulas predicted, with the result that the theory could be forced into apparent agreement with the data only by using values of *K* which increased indefinitely with distance.

About the same time the British mathematician L. F. Richardson (later to be celebrated as the founder of "numerical forecasting"[1]), who was not in government service and therefore had no knowledge of the Porton results, reached a similar conclusion by the examination of far less precise and accurate data. In the absence of more reliable information, he was forced to estimate diffusion from imprecise observations, including the results of a competition with toy balloons (a prize went to the child whose balloon covered the greatest distance before it burst) and the scatter of ash from volcanic eruptions, but some of his estimates referred to distances of travel far greater than those used at Porton. He found that to get agreement with the observed scatter the coefficient of diffusion had to be allowed to increase with distance

[1] See Chapter 8.

from the source, as in the Porton trials, but his distances ranged up to hundreds of miles and his values of the "constant" K varied from 10^2 to 10^{11} square centimeters per second!

It is, of course, impossible for a true physical constant to depend on distance from the source of the diffusing matter, for the position of the source is fixed by man, whereas the rate of diffusion at any point must depend entirely upon the nature of atmospheric processes in the neighborhood of the point and not on the distance of the point from an arbitrary origin. The only conclusion to be drawn from the diffusion trials is that no constant coefficient of diffusion exists, and that there is a fundamental difference between molecular and eddy diffusion which is more than a scale effect. This view is now accepted by meteorologists and the "constant diffusivity" theory, useful though it had been in the initial investigations of atmospheric turbulence, has long ceased to be a subject of serious research. A more flexible approach came into meteorology from aerodynamics.

The Mixing Length

In the kinetic theory of gases the average molecular velocity and the mean free path, the essential ingredients of all diffusion coefficients, are supposed to have the same value at all points in the field of motion. If this concept is carried over rigidly to the turbulence problem, the implication is that all atmospheric eddies are of much the same size and that on the average they travel the same distance before mixing (the eddy counterpart of "colliding") with other eddies. Yet even the most casual observation of smoke blown by the wind suggests that eddies tearing the cloud apart are of widely different sizes and that some persist for a long time before losing their identity, while others are remarkably evanescent. Further, it is clear that the proximity of the surface has a marked effect. Large discrete eddies cannot exist in very shallow layers near the ground.

Considerations of this kind led the German mathematician Ludwig Prandtl to propose a modification of the kinetic theory approach which allows the eddy counterpart of the free path to vary with position in the field of flow, but before considering this analysis we need to look more closely at the structure of turbulent flow.

If the record of wind speed near the ground is averaged over successive periods of many minutes, the result is the *mean velocity.* The difference between the mean velocity and the total velocity is the *eddy velocity.* In the relatively shallow layers with which micrometeorology is concerned, usually not more than 100 or 200 feet in depth, the deviating force of the Earth's rotation does not matter and the meteorologist is free to choose his system of axes without reference to compass directions. In the mathematical analysis of motion near the ground it is customary to take the x axis to be the direction of the mean wind, so that if u, v and w are the components of the velocity of the mean wind along, across and perpendicular to its direction, respectively, both v and w are zero. On the other hand, the components of the eddy velocity, which are usually denoted by u', v' and w', are not zero. They measure the downwind, cross-wind and vertical motions which are responsible for turbulent mixing. The quantities u'/u, v'/u and w'/u are known as the components of the *gustiness* of the wind and u'^2, v'^2 and w'^2 measure the *eddy energies* in the three directions.

A great deal of careful work has been directed to the exploration of the atmospheric eddy velocity field. It was soon realized that the oscillations have periods varying from a fraction of a second to many minutes, a feature of atmospheric turbulence which greatly complicates the mathematical analysis. In wind-tunnel investigations special devices such as honeycomb grids are used to break up the larger swirls, so that the eddies in the working section of the tunnel are always small and as uniform as possible. In the atmosphere no such simplification is possible. If we regard the oscillations as waves, we can speak of the *spectrum* of turbulence, and it is a matter of prime importance to the meteorologist to ascertain how the eddying energy is distributed over the spectrum. Investigations, not yet complete, indicate that much of the energy resides in the more rapid oscillations.

A further complication arises from the *partition* of eddy energy in the surface layers of the atmosphere. In the molecular theory it is assumed, with good reason, that on the average the three components of molecular energy are equal. Near the ground the lateral component of eddy energy (v'^2) is much greater than either the downwind or vertical components. In diffusion this is reflected in the fact that a smoke cloud generated at ground level spreads much more laterally than vertically in conditions of small temperature gradient. This feature of atmospheric turbulence was first demonstrated by Sir Geoffrey Taylor by the use of a tethered balloon, and later by the simple but ingenious instrument called the *bidirectional vane* or *bivane,* which is a very useful tool in field work. The bivane is simply a light wind vane with lateral and vertical fins mounted so that it can swing in both the horizontal and vertical planes. It thus provides a simple means of measuring v' and w' and was extensively used at Porton for this purpose. The instrument is arranged so that its movements are recorded by a pen on a drum. The record is a tangled mass of lines (Fig. 43) which, over periods in excess of 2 or 3 minutes, take a well-defined oval shape when the instrument is mounted near the ground, showing that at this height v' is significantly greater than w'. At heights in excess of 75 feet A. C. Best found that the components of eddy velocity are equal, the shape of the bivane record being very nearly a true circle.

Prandtl's approach arose from wind-tunnel investigations in which many of the complications described above do not occur. He wished to find a means of calculating the more important features of turbulent flow, such as the profile of the mean motion, the shearing stress, and the friction of the surface, in terms of readily observed properties of the mean flow. To do this he went back to the primitive concepts of the kinetic theory of gases and argued that if an eddy, which he still regarded as a small discrete volume of fluid, breaks away from an original level z and reaches another level $z + l$ before mixing, it will cause a typical fluctuation, or eddy velocity, the magnitude of which he was able to express in terms of l and the velocity gradient. He was then able to find a simple expression for the eddy viscosity which involved only l and the square of the velocity gradient. In Prandtl's theory the quantity l, which he called the mixing length (*Mischungsweg*), is a unique length characteristic of the local intensity of the turbulent motion but, unlike the free path of the molecular theory, can

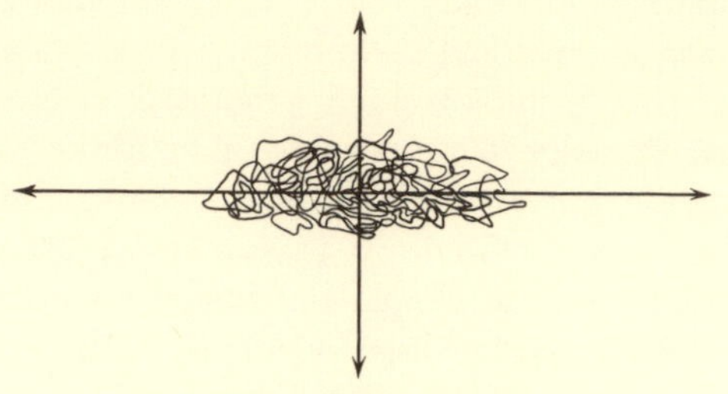

a. Record taken in lapse conditions about 10 inches above the ground

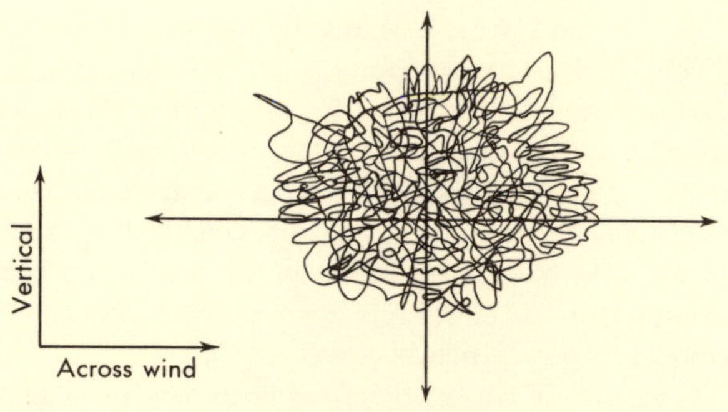

b. Record taken in lapse conditions about 20 feet above the ground

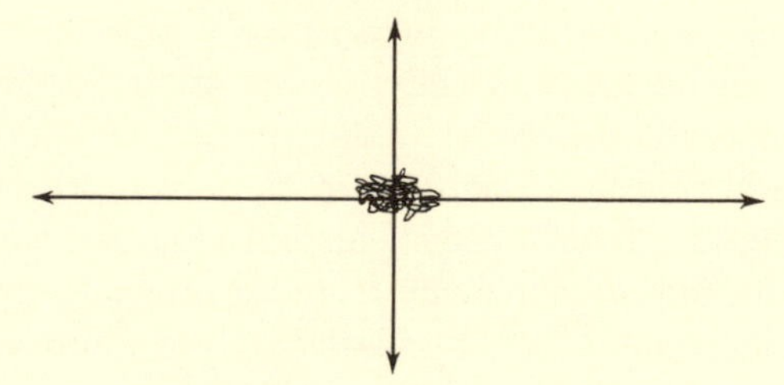

c. Record taken in inversion conditions about 6 feet above the ground

Fig. 43. Bivane records over level country.

vary in magnitude from point to point. In particular, it varies with height above the boundary. Such a variation is not open to logical objections like those which excluded the possibility of K varying with distance from the source in the smoke diffusion problem, for the surface of the earth is a real entity in the atmospheric turbulence problem, and not something introduced by man.

The expressions derived by Prandtl, when used in meteorological problems, enable the magnitude of the mixing length to be computed. It turns out that l is usually an appreciable fraction of the height of the point of observation above the ground and thus is to be reckoned in meters in most meteorological studies. The mixing length of atmospheric turbulence is much greater than its counterpart, the free path of molecular agitation, which is of the order of a hundred-thousandth part of a centimeter.

This simple but powerful idea has enabled realistic studies to be made of the variation of mean wind speed with height in the surface layers. The velocity profile near the ground varies considerably with the state of the motion. In the lapse period there is a vigorous exchange between the lower and upper air layers, and momentum is freely transferred in the vertical. The slower moving air near the ground is carried upward by the turbulence and is replaced by faster moving air from above, with the result that there is relatively little change of mean wind speed with height. In the inversion period, when turbulence is at a low level throughout the entire air mass moving over the surface, there is little exchange of the upper and lower air. In consequence, fresh momentum from above is not fed in sufficiently rapidly to replace that absorbed by the drag of the ground, and the mean wind speed drops to a low value very near the surface and varies rapidly with height. The profile of the mean wind is thus a sensitive indicator of turbulence, and consequently of the diffusion of momentum, in the surface layers.

Observations of mean wind speed near the ground, especially those made by the American meteorologists C. W. Thornthwaite and P. Kaser, show that when the sky is heavily overcast and the vertical gradient of temperature is small, the profile is accurately represented by the *logarithmic law of variation of wind with height,* which states that the mean speed at any level is proportional to the logarithm of the height of the level. This law, which holds only for relatively shallow layers near the ground, is valid also for flow near a solid boundary in a wind tunnel. Prandtl showed that the logarithmic law can be deduced by mathematical reasoning from the equations of turbulent motion if the mixing length is made proportional to height. Later, the Austrian mathematician Theodor von Kármán showed that a wide variety of wind-tunnel observations on flow near solid surfaces can be explained if the simple relation $l = kz$ is adopted, where k is a pure number known as *Kármán's constant* and z is distance from the surface. One of the most striking results of recent years is that found by Professor P. A. Sheppard, of London, that mean wind profiles and frictional effects observed near the ground in conditions of small temperature gradient are consistent with Kármán's formula with $k = 0.45$, which is close to the value 0.4 found in wind-tunnel experiments. The result is very remarkable when one considers the enormous difference of scale in the process of the transfer of momentum in the two cases.

The mixing length theory has thus gone a long way toward producing working formulas for the transfer of momentum near the ground in conditions when there is no marked density gradient in the vertical. In these circumstances many small and large-scale processes can be brought into line by one coherent theory. But the mixing length theory does not solve all problems, and we shall now leave this aspect and consider an entirely different type of approach which has had important consequences in the treatment of problems of the diffusion of matter. This is the statistical theory of turbulence, which originated in a highly original paper by Sir Geoffrey Taylor published in 1921.

The Statistical Theory of Turbulence

The theory of turbulence which we shall now describe depends largely upon the concept of *correlation* in statistics. In the course of a scientific inquiry it often happens that there are reasons for believing that changes in one quantity are either caused, or influenced, by changes in another quantity but the exact relation cannot be formulated. It may also happen that both quantities are influenced by a third variable quantity. The concept of correlation is an attempt by statisticians to measure closeness of relationship numerically. If we knew no dynamics and set out to find if the size of a small sphere affected its velocity of fall in air, we should find that if we measured the diameters and fall speeds of large numbers of different-sized spheres in identical conditions and plotted the results as a graph, the points would all be on a smooth curve, showing the speed of fall increasing with diameter. We could then say that the size of a sphere is correlated with its speed of fall. In this instance we know that an exact expression for the terminal velocity of a small sphere in terms of its diameter can be deduced by elementary dynamics from Stokes' expression for the resistance experienced by a sphere moving slowly through air, so that there is no need to call in statistical theory, but in many problems an exact relationship cannot be found. This frequently happens in meteorology. For example, there may or may not be a relation between rainfall and numbers of sunspots. We cannot hope to discover an explicit connection, expressible in a mathematical formula, at the present stage in the development of atmospheric physics. All that can be done is to examine the data in the light of statistical theory to see if the weight of evidence is in favor of a real connection between the two phenomena.

A correlation coefficient is a number which measures the closeness of the relation between two quantities which vary at the same time. The coefficient can take any value between $+1$ and -1, including zero, the calculation being carried out by an arithmetical process which is entirely objective. A coefficient of $+1$ between two sets of numbers means that they are perfectly related and that, on the average, they increase or decrease together. The value -1 means that there is again a perfect relation but that one set decreases as the other increases. A zero correlation coefficient implies that there is no relation between the numbers.

One of the earliest demonstrations of the reality of the fundamental tenets of the kinetic theory was afforded by the phenomenon known as *Brownian motion,* first noticed in 1827 by the botanist Robert Brown when examining a suspension of particles of pollen under the microscope. The particles wander in a random fashion across the field of view. This motion is caused by the impacts of molecules on the tiny spheres. The individual motions are extremely irregular, but they exhibit statistical regularity. This was shown by Einstein, who proved mathematically that the average distance traveled by a particle is proportional to the square root of the time taken. Einstein's formula was afterward verified by the French scientist Perrin, who measured the scatter of large numbers of particles.

Brownian motion is related to a famous statistical problem known as the *drunkard's,* or *random, walk.* A molecule in a gas experiences a very large number of collisions in a very short time,[2] which means that it must follow a very irregular path. Suppose, for simplicity, that all the free paths are of the same length l, and that the molecule always moves in the same direction, but sometimes backward and sometimes forward. In a given time the molecule will cover a distance $l \pm l \pm l \pm \cdots \pm l$, and because we have no way of knowing when a forward or a backward step will occur, it might seem hopeless to try to discover any laws for such random motions. This must be true when the number of steps is small but if, as in Brownian motion, the number of impacts is enormous even in relatively short intervals of time, it turns out that a simple law connecting the *average* distance traveled by the particle and the time taken can be deduced. When the steps are distributed at random, the distance traveled by the particle in a given time must, on the average, be less than the distance covered by a particle subject to nonrandom impacts, and the law in question is that for random motion the *average* distance traveled is proportional to the *square root* of the time that has elapsed since the observations began. If this law is applied to an inebriated citizen, we can say that on the average he takes four times as many steps to travel two miles as to travel one—a good argument for sobriety when walking.

In Brownian motion, and in molecular motions in general, it seems that the motion is completely random in the sense that there is no correlation between successive steps. Does anything like this happen when a smoke cloud is scattered by wind eddies? Atmospheric diffusion is similar in many ways to a random walk process, but there are differences. On looking at a smoke cloud drifting in

[2] Of the order of a thousand million a second in normal conditions.

the wind the impression gained is that a cluster of particles starts to spread under the influence of a fairly well-defined system, but that as the cluster becomes larger it is increasingly influenced by the motions of other systems which bear no obvious relation to the original system affecting the particles. When the particles are close together, eddies of diameter much greater than the average distance between the particles move them as a mass and do not separate them. Separation of the particles is caused only by eddies smaller in size than the average distance between the particles, and as the cloud grows, and the particles drift farther and farther apart, bigger and bigger eddies begin to play a part in the process of separation. In mathematical language, at the beginning of the diffusion process the correlation between the initial and subsequent motions of a typical particle is high, but as time proceeds the relation grows weaker and must ultimately become negligible. In terms of the drunkard's walk, it is as if the subject set off in a definite direction and with a set purpose, but that as time goes on his route bears less and less resemblance to that originally planned.

In 1921 Taylor studied an extension of the random walk problem in which there is a variable correlation between the motion of a particle at one time and at some instant later. He discovered a simple equation which expresses the scatter of a cluster of particles in terms of the correlation coefficient. This equation is now one of the most celebrated in fluid mechanics and has been the starting point for numerous investigations in recent years, but at the time it was published it did not attract much attention among workers in fluid-motion theory, possibly because it appeared in a journal devoted to pure mathematics but more probably because the ideas on which it was based were in advance of their time.

The relation between the motion of a particle at one time and at some instant later is called *Lagrangian correlation,* the reference being to the system of hydrodynamics invented by the great French mathematician Joseph-Louis Lagrange (1736–1813), in which attention is concentrated on the behavior of a single particle throughout the motion instead of on the velocity field of all the particles, as is done in the system invented by Euler. It is not practicable to measure the Lagrangian correlation coefficient directly in the atmosphere, and there does not seem to be any way in which the functional form of the coefficient can be deduced by strict mathematical reasoning, except in very special circumstances like that of Brownian motion, which approximates closely to a series of disconnected jerks. In Brownian motion the correlation coefficient is effectively zero except for a very short time at the start of the motion. When this behavior of the correlation coefficient is expressed in mathematical form and used with Taylor's equation the result is equivalent to Einstein's formula for the scatter of microscopic particles by molecular impacts.

Some measure of empiricism, like that employed in the mixing length theory, seems therefore to be necessary if the ideas inherent in Taylor's equation are to be incorporated in diffusion theory. About 1930 the writer, who was then a member of the Porton group of meteorologists, suggested that as a smoke cloud expands by diffusion it comes progressively under the influence of larger and larger eddies and that at any stage in its growth there is likely to be a dominant size of eddy, but nothing in the way of a definite model of eddy motion was proposed. This indicated that the Lagrangian correlation coefficient might be expressed in a relatively simple form, in which the closeness of the relationship between the initial and subsequent motions separating the particles is made to decrease at a definite rate. If t (time) is the interval of diffusion, the simplest hypothesis that can be adopted is that the correlation coefficient behaves like t^{-n}, where n is a positive number. It is then a simple matter to use Taylor's equation to find an expression for the size of the diffusing cluster at time t after it was formed, in terms of the wind speed, the eddy energy, and the quantity n. The problem then becomes that of finding how to determine n from meteorological observations.

It has already been shown[3] that the rate of increase of the mean speed of the wind in the lowest layers of the atmosphere is a sensitive indicator of the degree of turbulence in the atmosphere. The wind profile, in fact, shows how effectively the eddies spread momentum from one level to another. Common sense suggests that there is no essential difference between the diffusion of momentum and the diffusion of matter by turbulence, for the basic mechanism must be the same in both phenomena. When the theory

[3] See p. 151.

outlined above was applied to the calculation of the velocity profile near the ground, a simple means was found of evaluating the elusive quantity n. In conditions of small temperature gradient (overcast skies and moderate or high winds) the average value of n is ¼: in inversion conditions n increases but never reaches 1, and in the lapse periods n is less than ¼ but never reaches zero.

The crucial test of this work lay in comparisons with measurements of diffusion made in the studies of the spread of smoke and gas clouds. Roberts' earlier work had shown that the "constant diffusivity" hypothesis led to the result that the central concentration in a continuous point-source cloud should decrease as x^{-1}, where x is distance downwind, whereas measurements made in conditions of small temperature gradient indicated that the true law was close to $x^{-1.76}$. In the theory described above the corresponding result is x^{n-2} which, for $n = ¼$, the value deduced from observations on the mean wind profile, becomes $x^{-1.75}$ in almost exact agreement with the experimental results. The results for the fall-off of concentration from long-line source clouds were equally satisfactory. Further, the theory predicted that the rate of decrease of concentration would be much slower in inversion conditions because of the increased values of n, and that in the limit, when turbulence is completely extinguished, the fall-off would be as x^{-1}. This limit appears never to be reached in the atmosphere, where there is always some residual turbulence.

In its original form the theory outlined above was satisfactory as regards the functional form of the expressions for concentration but less accurate in its predictions of the *scale* of atmospheric diffusion. Later a close measure of agreement with observations was obtained by making allowance for the fact that atmospheric motion is of the type known in aerodynamics as *fully-rough flow*. If the boundary surface is very smooth, turbulent motion becomes laminar in a thin layer at the surface itself. In this "laminar sublayer" the kinematic viscosity is important. However, if the surface contains protuberances which are large enough to pierce the laminar layer and so prevent its formation, the flow becomes of the fully-rough type, in which the molecular constants play no significant part. Investigations have shown that nearly all natural surfaces are "aerodynamically rough," and this has important effects on wind structure and the diffusion of matter.

The mathematical study of atmospheric turbulence has now become a matter of considerable interest to meteorologists. The latest advances involve many abstruse concepts and require a deep knowledge of mathematics for their understanding. For that reason only an outline of the more recent developments will be given here. In 1935 Taylor greatly extended his work on the statistical theory of turbulence and made some far-reaching investigations into the spectrum of turbulence. In 1941 the Russian mathematician A. N. Kolmogoroff advanced some novel and significant suggestions concerning the transfer of energy by turbulence, which have already been referred to in Chapter 4. This work was in some respects similar to that of other mathematicians, notably Heisenberg, Onsager and Weiszäcker. To some extent these theories had been anticipated by one proposed in 1926 by L. F. Richardson, which attracted little attention at the time, possibly because it was written in a somewhat obscure style. Richardson's ideas may be summarized thus: In any diffusion process the distance between two marked particles will increase as time proceeds. Molecular diffusion (which is sometimes referred to as *Fickian diffusion*) is characterized by the feature that, on the average, the rate of separation of two such particles is independent of the distance between them. Richardson, by a mathematical process which he called the *distance-neighbor graph,* studied a more general type of diffusion in which the rate of separation of two particles increases with their distance apart. In 1947 he enlisted the aid of the American oceanographer Henry Stommel to try to obtain an experimental verification of his deductions. It was characteristic of Richardson's gift for getting evidence from unlikely sources that one test consisted in observing the behavior of parsnips floating on a lake in Scotland. The result, one is glad to say, was not unsatisfactory and showed coherence with earlier data for the atmosphere.

In this chapter we have touched upon the salient features in the history of the development of what is now a busy sector of meteorological research. In the next chapter we shall examine how some of this work has found practical application.

CHAPTER 7

MICROMETEOROLOGY AT WORK

The work of meteorologists is only too often represented to the world simply as a kind of insurance against the evil effects of weather, and it is undeniable that many judge the success of a national meteorological service mainly by the number of occasions on which warnings of impending storms have been justified by subsequent events. There is, however, a positive side to the picture which is well exemplified by micrometeorology, and we shall give examples to show that the difficult and seemingly academic mathematical studies described in the previous chapter are proving of benefit to mankind in many different ways.

Atmospheric Pollution

In the seventeenth and eighteenth centuries the disposal of sewage in the rapidly growing cities of Europe was carried out in a manner which today seems unthinkable. It may well be that our descendants will look back with equal incredulity to the nineteenth and the early-twentieth century as a time when people threw noxious and evil-smelling substances into the atmosphere with little or no regard to life and property. Today the urgency of the problem of atmospheric pollution is recognized by all, if only because of the advent of nuclear energy. The real remedy is to prevent the escape into the air of all substances injurious to health, but in an industrial civilization this is at present a counsel of perfection. Meteorology can help in this problem by providing reliable estimates of the concentrations of pollution likely to be found in the vicinity of a source as an aid to planning, and it is mainly for this reason that in recent years much attention has been paid to the study of the spread of airborne matter, both gaseous and particulate, from industrial stacks.

When smoke or gas is emitted from a high stack, any massive particles (such as grit) contained in the effluent quickly fall out and are deposited near the source. Most industrial plants have installed arresters to remove such particles before they reach the orifice, and here we shall assume that the effluent is either entirely gaseous or contains only particles small enough to have a negligible terminal velocity and are thus airborne throughout the entire travel of the cloud. It has been shown in many trials that such "particulate" clouds diffuse in a turbulent wind at the same rate as a gaseous cloud. We assume also at this stage that the effluent has the same density as the ambient air.

When there is a high lapse rate of temperature in the lower layers of the atmosphere, the plume of smoke from a stack follows a highly irregular path characterized by "looping," or large wave-like motions (Fig. 44, a). When the temperature gradient is small and the wind moderate or strong, the smoke stretches downwind in a horizontal cone (Fig. 44, b). Finally, in marked inversion conditions the smoke is concentrated into a narrow cylinder which may travel for many miles with very little widening (Fig. 44, c). All these changes are manifestations of the control of turbulence by temperature gradient and are reflected in the character of the wind-direction records.

When the source of the smoke or gas is at ground level, the surface of the earth forms an impermeable barrier which may be regarded as dividing the cone of the cloud in half. As a result, concentrations in a ground-level cloud are everywhere double those that would be obtained from the same source in the same conditions of turbulence, but generated far away from a rigid barrier. In this sense the ground acts as a "reflector," but this does not mean that the smoke literally bounces off the ground as if the particles were elastic spheres. The term "reflector" is here simply a mathematical term which describes the effect of an impermeable barrier.

When a cloud is generated at a considerable height above the ground, the effect of the surface is more complicated. Figure 45 shows an idealized representation of ground-level concentration in such a cloud. At first the smoke is blown away from the orifice and travels as if the ground did not exist, but as the cloud deepens with distance downwind a stage is reached when the smoke comes into contact with the surface and is "reflected." An observer walking downwind from the foot of the stack would not detect any effluent at first, but as his distance from the stack increased he would find that the concentration rose, reached a maximum, and

a. Large lapse rate, moderate wind

b. Small lapse rate, moderate or high wind

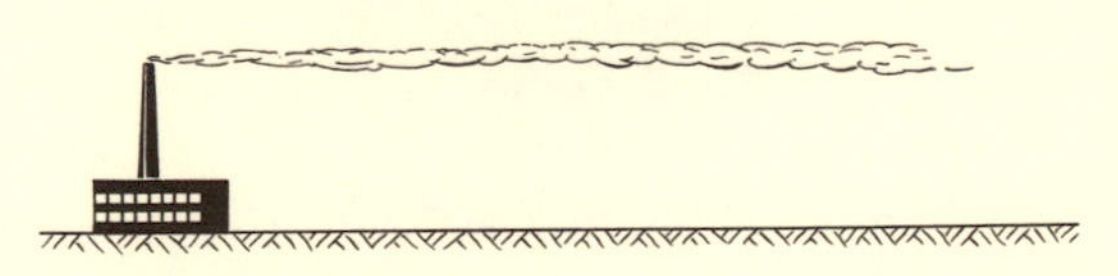

c. Large inversion, low wind

Fig. 44. Behavior of smoke from a stack in different conditions of temperature gradient.

then decreased rather slowly. This variation of concentration with distance is quite unlike the continuous decrease observed in a cloud from a similar source at ground level.

The task of the micrometeorologist is to calculate from meteorological observations the concentration at ground level downwind of the stack. The mathematical theory can deal only with concentrations averaged over many minutes, for it is clearly impossible to

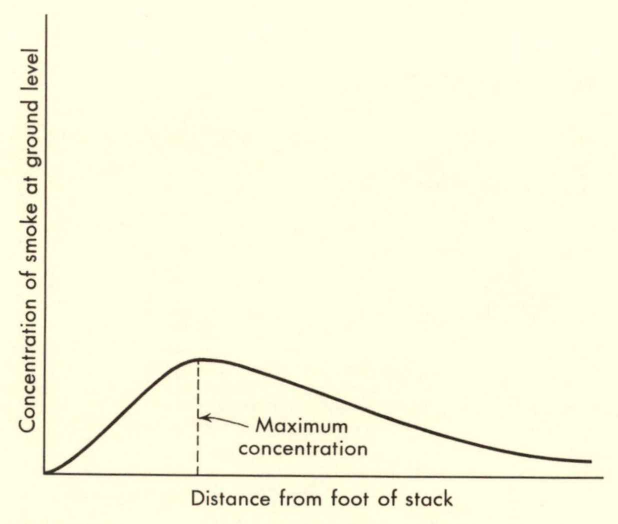

Fig. 45. Concentrations in a smoke cloud from a stack.

take account of every fluctuation of concentration, especially in conditions of steep lapse rate which produce "looping." Nor are such instantaneous values required for the purpose of deciding the hazard to health liable to arise from the operation of an industrial plant. Account must also be taken of the fact that the turbulent energy of the wind decreases with height, and it is necessary to employ mean values of the diffusion coefficients appropriate to the whole of the layer in which the smoke diffuses.

The mathematical technique employed is known as the *method of images*. It is supposed that below the line which represents the impermeable surface of the earth there is a second, hypothetical, source identical in all respects with the real source, which produces a cloud which is the "mirror image" of the real cloud in the line representing the ground. In this system there is no net transfer of matter across the ground-level line (Fig. 46). For every particle that crosses the line in one direction a similar particle moves in the opposite direction from the mirror-image cloud. In this arrangement the line representing the ground has thus all the properties of an impermeable surface, and to solve the mathematical problem all that has to be done is to set up such a system with the appropriate diffusion theory and to combine the solutions for the two sources.

Fig. 46. The method of images.

This method was used by the writer in 1947. The main result is that if Q is the strength of the source (the rate at which smoke or gas leaves the stack orifice), h is the height of the stack orifice aboveground, and u the mean wind speed in the layer extending from the ground to the top of the stack, the maximum concentration of smoke at ground level is

$$\frac{2Q}{e\pi uh^2}$$

where e is the base of natural logarithms ($e = 2.72$ approximately). The distance downwind from the foot of the stack at which this concentration is found is

$$\left(\frac{h}{C}\right)^{2/(2-n)}$$

where n is the number which expresses the degree of turbulence (see Chapter 5) and C is a generalized diffusion coefficient which can be evaluated from meteorological observations.

These results show, first, that the maximum concentration at ground level varies inversely as the square of the height of the stack. This law, which had been found by other methods in 1936 by C. H. Bosanquet and J. L. Pearson, shows precisely the value of high stacks in the dispersal of atmospheric pollution. Doubling the height of the stack, for instance, reduces ground-level concentrations by a factor of four. Second, the distance of the point of maximum concentration from the foot of the stack increases very nearly in proportion to the height of the stack in conditions of small temperature gradient, and more rapidly in inversion conditions. The theoretical effect of an inversion is to move the concentration curve downwind, leaving the maximum concentration unchanged, but this result may arise in part from the simplified theory employed and it is not known how far this deduction can be relied upon in practice. The nocturnal inversion affords relief to those living close to a source of pollution because the cloud is kept aloft for considerable distances, but it also has the adverse effect of increasing concentrations far from the vicinity of the source. A special nuisance arises when the cloud which has drifted a long distance during the night without touching the surface breaks up and diffuses to ground a few hours after sunrise because of the generation of turbulence in the lower layers of the atmosphere by solar heating. This phenomenon is sometimes called *fumigation*.

In this treatment it is supposed that the effluent has the same density as the surrounding air, so that there are no buoyancy effects. In practice, most industrial gases are much above air temperature at the point of release, and this has a beneficial effect on concentrations of pollution near the source. The same effect can be produced, but less efficiently, by a powerful forced draft in the stack. In both instances the result is that the cloud ascends well above the level of the orifice and the effective height of the stack is increased, causing a marked reduction in ground-level concentrations of pollution.

The thermal effect was particularly well shown in a pollution problem which arose in connection with a smelter at Murray, Utah. A stream of gas containing sulfur dioxide from the smelter had caused damage to vegetation in the vicinity of the plant.[1] The main source of pollution was a stack 200 feet high. Some relief was gained when the height of the orifice was increased to 300 feet, which by the inverse-square law between stack height and concentration would bring about nearly a 50 per cent decrease in ground-level concentrations. An attempt was then made to achieve a further reduction in pollution by blowing in cold air at the foot of the stack. This had an adverse effect, for the incoming air cooled the gases in the stack and increased concentrations of sulfur dioxide were measured at ground level between a quarter and a half mile downwind of the plant. Finally, a stack heater was introduced and the height of the orifice increased to 450 feet, after which the concentrations were reduced to an acceptably low level. This happy outcome can be attributed mainly to the rise of the heated plume.

The problem of the path taken by a plume of gas from a hot source is thus of more than academic interest. It is not difficult to investigate the dynamics of such a plume in calm air in neutral static equilibrium (i.e., an atmosphere in which the temperature of the ambient air decreases with height at the adiabatic lapse rate) and solutions which have been verified in the laboratory have been found by Schmidt (1941) and the writer (1950). The problem is much more difficult in an atmosphere in which there is an inversion of temperature, but even in this case considerable progress has been made. The height reached by the ascending smoke depends not only on the intensity of the heat source but also on the distribution of ambient air temperature with height. It is a common sight at sea to see the smoke from a steamship rise rapidly at first and then spread out horizontally at the inversion top. In a calm atmosphere in which temperature falls with height at the average rate observed in the real atmosphere it has been calculated that the effluent from a large power station, in which about ½ megakilowatt goes up the stack, rises to nearly 4,000 feet. In the same conditions the smoke from an open household fire, burning

[1] G. R. Hill, M. D. Thomas, and J. N. Abersold, *Proc. 9th Annual Meeting,* Industrial Hygiene Foundation of America, 1944.

about 4 pounds of coal an hour, with half the heat wasted up the chimney, ascends to about 250 feet.

The most serious mathematical difficulties arise in the problem of a heated plume in a horizontal wind, and there are many different solutions from which to choose. A simple empirical expression, which takes into account both thermal and forced-draft effects, is that derived by the American meteorologist J. Z. Holland from observations made at Oak Ridge, Tennessee.[2] Holland found that a heat source of Q calories a second with a forced draft of v miles an hour causes the plume from an orifice of d feet diameter to rise

$$\frac{1.5\,vd + 10^{-4}Q}{u}\text{ feet}$$

above the top of the stack in a wind of u miles an hour.

Holland's formula is entirely empirical and bears no relation to the various theoretical expressions which, in turn, bear little resemblance to each other. Calculations show, however, that despite these apparent differences all the formulas lead to much the same values for the reduction of concentration at the ground arising from the buoyancy of the smoke. There now seems to be good reason for saying that useful approximate solutions have been found for a problem which is bound to grow in importance as more and more of the world's supply of energy is derived from fission processes. The golden rule for designers of industrial plants which emit pollution is: *Keep the stack orifice high and the gases hot.*

There is, however, a further matter to be considered by the industrial architect. The meteorological studies assume that the effluent gets away from the stack orifice without hindrance. This is not always the case, and a cliff-like building can produce a substantial lee eddy which forces down the plume in the vicinity of the source. It is now common practice to look for such effects, before the building is constructed, by wind-tunnel investigations with accurately scaled-down models, since an exact calculation of the flow is rarely possible. But it is not possible to reproduce in a wind tunnel the temperature gradients which play such a major

[2] Atomic Energy Commission Report ORO-99 (1953).

role in the control of atmospheric turbulence, and for an estimate of such effects it is necessary to look to theory.

Another effect, well recognized by meteorologists, which also is not, in general, amenable to mathematical analysis, is that of topography. If an inversion occurs over a valley in an industrial area, the result may be especially serious, for the sides of the valley act as additional barriers to prevent the escape of polluted air. In our own time there have been two outstanding examples of such effects, in the Meuse Valley of Belgium in December, 1930, and at Donora, Pennsylvania, in the valley of the Monongahela, in October, 1948. The great London smog of December, 1952, cannot be attributed to topographic influences, and it is still not clear why this particular combination of pollution and fog was so deadly. Nor is the relation between pollution and fog established beyond doubt, for the meteorological conditions which inhibit diffusion are precisely those which favor radiation fog. It seems likely that the main effect of pollution is to make fogs denser and more persistent, and, of course, more dangerous to health, and for these reasons alone the study of atmospheric pollution is amply justified.

Evaporation

The conduction of heat and the evaporation and condensation of water are the fundamental energy transfer processes in the atmosphere. The problem of evaporation, which is especially important in hydrology, may be approached in two, somewhat different, ways. In the first method the objective is to calculate the rate at which water is removed from a surface by the wind using the theory of eddy diffusion. In the second method the details of the transfer process are not specified, and the process is studied as part of a general exchange of energy. The two methods are known as the *diffusion* and *energy-balance* theories, respectively.

It is well known that wet clothes dry more rapidly in the open on a warm windy day than on a cold calm day, and also that the rate of removal of water from the fabric depends very much on the humidity of the air. These simple facts provide the basis of all theories of evaporation. High air temperature means that the saturation vapor pressure of water also is high and, if the air is fairly dry, the *saturation deficit,* or difference between the actual vapor pressure of the water in the air and its saturation value, is large. The first empirical law of evaporation is therefore that the rate of removal of water from a wet surface is proportional to the saturation deficit, which for this reason is sometimes called (rather misleadingly) the *drying power of the air.* As the water-vapor molecules leave the surface they increase the vapor content of the air and thus reduce the saturation deficit. If this state of affairs continued indefinitely, the water vapor in the air over the surface would soon reach saturation and evaporation would cease. The wind sweeps away the vapor-enriched air, and clearly the stronger the wind the more efficient is the removal process. One of the earliest quantitative statements about evaporation was made in the early part of the nineteenth century by the English chemist John Dalton: that the rate of evaporation is proportional to the wind speed and the saturation deficit. This statement is often referred to as *Dalton's law.* If it expressed the whole truth, the calculation of evaporation from, say, a lake would involve no more than the determination of the constant of proportionality between the rate of evaporation and the product of the wind speed and the saturation deficit, for the total amount of water removed would be proportional to the area of the water surface. However, this is not always true, and the problem must be examined on more sophisticated lines.

It is necessary to distinguish between two types of problems in evaporation. The first, known as the *limited-area problem,* is concerned with the rate of removal of water as vapor from, say, a lake or a reservoir surrounded by dry land. Air with a vapor content below saturation comes from the land to the lake, and in so doing forms a blanket of vapor the depth of which increases with distance from the shore (Fig. 47). The rate of evaporation depends on the number of molecules that leave the surface of the water less the number that return, and this number clearly depends in turn upon the concentration of water in the blanket of vapor above the surface. In this problem we must therefore expect the rate of evaporation to decrease with distance from the upwind edge of the lake, and this in turn means that the total amount of water evaporated is not simply proportional to the area exposed.

In 1934 the writer solved the "semi-infinite area" problem of evaporation in terms of the theory of diffusion which had been developed from Taylor's "drunkard's walk" theorem of 1921 (see Chapter 6). In the semi-infinite area case the evaporating surface is supposed to be very long across wind but of finite length down-

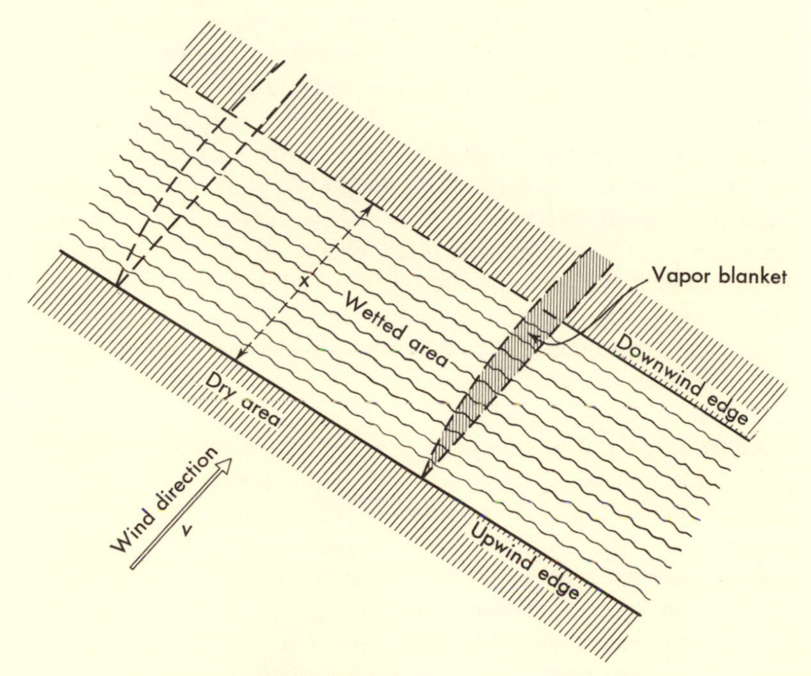

Fig. 47. The semi-infinite area problem of evaporation.

wind. (This is simply a mathematical way of saying that the sideways transfer of vapor is neglected.) The problem cannot be solved without specifying the concentration of vapor at the surface of the water, and the assumption was made that this was always equal to the saturation value.

The complicated mathematical formulas for the rate of evaporation derived in this way have been confirmed by wind-tunnel experiments made in many different laboratories. They show that the simple Dalton's law is nearly true as regards the effect of wind speed. The problem of evaporation from an area which is of finite dimensions across wind as well as downwind is far more difficult and has not yet been solved.

The second type of evaporation problem is that which arises when there is no abrupt transition from a dry to a wet surface and therefore no boundaries to be considered. This is so, for example, in the problem of evaporation from the ocean. The direction of the wind clearly does not matter and the rate of evaporation is the same at all points. For this "infinite-area problem," one method employed is to measure the vertical gradients of wind speed and humidity in a shallow layer near the surface. From such measurements it is possible to estimate the flux of water vapor from the ground. Methods of this kind, which call for extremely accurate instrumentation, are used chiefly to assist the study of horticulture. They give the rate of evaporation from soil and from crops only in the immediate vicinity of the measuring apparatus, and usually this is all that is needed to assess the water budget of a crop at an experimental farm.

This method is laborious and needlessly precise for the purposes of climatology and hydrology. For very large areas the assessment of evaporation is carried out by the simpler energy-balance method. Because of the large latent heat of water the process of evaporation, whether it be from a lake, from bare soil or from plants, requires a considerable amount of energy, which must come from solar radiation, either direct or diffuse. The solar energy undergoes many changes: some of it is reflected, and some emitted by the surface as long-wave radiation, some is used to warm the air above the surface, and the body of the water or the soil, and some provides the latent heat of evaporation. Climatological studies enable estimates to be made of the net income of radiation and of the flux of energy from the surface as sensible heat, and studies in soil physics provide an estimate of the amount of heat which passes into the soil and is stored. Since energy is conserved, it is possible to estimate the heat used in evaporation as a difference and hence to calculate the actual rate of evaporation.

Let

F_R = the net radiation income of a surface
F_H = the flux of energy from the surface as sensible heat
F_S = the flux of heat into and stored by the soil
F_E = the heat used in evaporation.

By the principle of conservation of energy, the income is shared between the last three items or, in the form of an equation

$$F_R = F_H - F_S + F_E$$

(net energy income) = (warming) + (heat used in evaporation) and thus F_E, and hence the rate of evaporation, can be found from estimates of F_R, F_H and F_S.

This method has been extensively used by workers in agricultural science to estimate how much energy is available to convert water into vapor through the leaves of a plant, a quantity called the *potential transpiration.* (On the average, in an English summer, about 40 per cent of the energy is available in this way.) It is thus possible to form an idea of the maximum water needs of a crop in a region for which reliable climatological records exist, and hence to determine the amount of irrigation, if any, that is required. The method represents a departure from the normal practices of micrometeorology in that no appeal is made to the theory of diffusion.

The water supply of a country is one of its most precious assets, and it is now recognized that a reliable assessment of the water budget of region is a matter of prime importance to a community. Every year sees an increase in the demand for potable water, and even traditionally wet countries like Britain may be forced in the future to conserve their water supply. Meteorology is an integral part of hydrology.

Night Minimum Temperatures

In Chapter 2 we already referred to the problem of the prediction of night frosts in a California valley as an example of the reduction of radiation loss by water vapor in the air. The precise determination of the radiational cooling of the ground at night is important not only for farmers but also in connection with the prediction of fog. In consequence there have been many attempts to produce concise formulas which help the forecaster to fix the level to which the thermometer will fall at night. Such formulas are empirical and usually incorporate "constants" which vary with locality and act as "corrections" for the effect of wind. A typical formula used in England is that devised by C. J. Boyden,

$$T_{\text{min}} = \tfrac{1}{2}(T_w + T_d) + C$$

where T_{min} is the night minimum temperature, T_w is the wet-bulb temperature and T_d the dew point at the time of the day maximum temperature. The quantity C is the "correction factor" for wind speed and cloud amount, which varies with locality.

A more sophisticated method of determining night minimum temperatures has been devised by W. E. Saunders. This depends upon the observation that the rate of cooling of surface air after sunset shows a discontinuity which is especially marked on clear nights with little wind. This sudden reduction in the rate of fall of temperature appears to be associated with the release of latent heat when dew forms on the grass. Suppose that t_r is the time at which the reduction begins (in England this is usually about an hour after sunset), T_m the maximum temperature of the air at 4 feet above the surface (the screen maximum) during the day, and T_d the dew point at the time of maximum screen temperature. Analysis of the records shows that T_r, the screen temperature at time t_r, is given with tolerable accuracy by the equation

$$T_r = \tfrac{1}{2}(T_m + T_d) - K$$

where

$K = 1°$ F if there is no inversion below 900 mb (about 3,000 feet)
$\quad = 4°$ F if there is an inversion

From the calculated value of T_r the subsequent cooling of the air, and hence the night minimum temperature, can be estimated from graphs. The Saunders technique has been found useful for the prediction of low-level fog at airfields.

For ground frosts R. Faust, following investigations at Münster, Germany, has proposed an empirical forecasting rule which says that if less than 2/10 of the sky is covered with cloud and the wind speed in the surface layers is not more than 4 knots, frost will occur at night if

$$T(14) + \tfrac{1}{2}T_d(14) < 79^\circ \text{ F}$$

where $T(14)$ and $T_d(14)$ are the screen temperature and dew point, respectively, at 1,400 hours local time. This rule has been tried at stations in England and found to be reasonably trustworthy.

The formulas quoted above are like old-fashioned pharmaceutical prescriptions in that they are made up of a number of ingredients which in combination may produce the required effect but there is little in the way of a clearly defined underlying theory. The use of air temperatures at about the time of the maximum determines the general level of temperature, and the dew point is included because of the controlling influence of the water vapor in the air on the loss of heat by radiation. The "constants" in the formulas indicate the influence of wind and of the physical properties of the soil.

A theoretical approach to the problem must take into account many factors which, however, are not independent. If the air were completely dry, the radiative loss would be determined by Stefan's fourth-power law (p. 11), but in practice this loss is largely compensated by the back radiation from the water vapor (and, to a lesser extent, the carbon dioxide) of the atmosphere. In addition, there is a flow of heat into or from the ground and, if a wind is blowing, a transfer of heat to and from the air by turbulence. However, a remarkably simple way of treating the problem was devised by Sir David Brunt in 1932.

From an analysis of observations on the backflow of radiation from the water vapor of the atmosphere, Brunt found that the flux could be represented with adequate accuracy by a simple expression which involved only the temperature and water-vapor content of the surface layers. In symbols, if T is the absolute temperature (° K) of the air near the ground and e is the vapor pressure in the same layer in millibars, the back radiation is given by

$$\sigma T^4 (a + b\sqrt{e})$$

where a and b are constants (average values are 0.44 and 0.08, respectively), and σ is Stefan's constant (p. 11). It follows that the net rate of loss of heat from the surface, R_N, is given by

$$R_N = \sigma T^4 - \sigma T^4(a + b\sqrt{e})$$

The variations in the absolute temperature and the absolute humidity of the air during the night are not large, so that this flux of energy may be considered to be constant, with little error. This assumption effectively eliminates the atmosphere from the problem, so that the temperature of the surface is decided by the condition that the flow of heat in the soil adjusts itself so that a constant flux is maintained across the surface. The problem is then reduced to one familiar in the mathematical theory of the conduction of heat in solids, namely, to determine the temperature of a body which is initially at a uniform temperature throughout but which experiences a steady flow of heat at a known rate across its surface. The solution, when expressed in meteorological terms, is *Brunt's parabolic formula for nocturnal surface temperatures,*

$$T(t) = T(o) - \frac{2R_N}{\rho C}\sqrt{\frac{t}{\pi\kappa}}$$

where

$T(t)$ is the temperature of the surface at time t after sunset
$T(o)$ is the temperature at sunset
ρ is the density of the soil, c its specific heat and
κ its thermometric conductivity.

Thus temperature at night decreases as the square root of the time since the supply of solar radiation ceased, and the graph of temperature as a function of time is a parabola.

Later investigations have followed this lead and theoretical formulas have been produced which include the effects of wind and of the release of latent heat when dew forms. Such solutions are necessarily more elaborate than the original parabolic law, but the gain in accuracy is not marked. The formulas yield good results only when the physical constants of the soil are known with considerable accuracy, and this makes them, in general, unsuitable for routine forecasting practice. When the density, specific heat and conductivity of the soil are known with precision, the results are impressive. Brunt, using the original parabolic formula, calculated a fall of 11° C for a dry summer night in England, as against a measured fall of screen temperature[3] of 9° C. The actual surface temperature was not measured, but it must have been somewhat lower than the screen temperature, so that the agreement undoubtedly was very good.

The Climate of the Lower Layers

Brunt's work shows clearly that changes of surface temperature are greatest for soil which has the lowest values of c, the specific heat, and κ, the thermometric conductivity. Low values of these constants occur with soil which contains much air space between the particles, such as sandy soil. A region with well-drained clay soil is therefore likely to have warmer nights and cooler afternoons, on the average, than one with a light sandy soil, and this may be an important factor in deciding the site of a house or a farm.

The temperature of the actual surface of the Earth is not easily measured continuously, but it is well established that very high values are reached in bright sunshine. The main difficulty of measurement arises from the fact that the rate of change of temperature is very large in the immediate vicinity of the surface. The most reliable values are those of the temperature of the soil just below the true surface. In desert soil at Tucson, Arizona, 161° F was recorded at 4 millimeters below the surface, so that the true surface temperature must have been far higher, and it is possible that surfaces exposed to the blazing tropical sun attain temperatures not far below 200° F. Even in England, a road surface has been found to reach 140° F, and a sandy surface 130° F, on a warm summer day.

These observations show that the climate in which a small ground insect lives, or into which a seedling emerges, is quite different from that recorded at screen level (4 feet). The highest screen temperature ever recorded is 136.4° F at Azizia in Libya, but it is clear that such temperatures are part of the daily experience of a ground insect in summer even in the temperate climates.

[3] See Appendix I.

Further, the diurnal range of surface temperature is enormous—in the Tucson example the range exceeded 100° F, and 62° F has been measured on a sandy heath in Finland. The physiology of small insects and other creatures living very near the ground must therefore be adapted to endure extremes and changes of temperature far greater than those experienced by larger animals.

The motion of the air in the surface layers is often very different from that at greater heights, and local winds may have little connection with the pressure pattern shown on the synoptic chart. Two types of local wind are especially important. The *sea breeze* is a shallow local coastal current caused by the higher temperature of the surface of the land in daytime compared with that of the adjacent sea. There is a greater expansion of a column of air over the land than over the water and a thermal circulation is set up from sea to land in the lower layers, with a return flow aloft. At night the direction is reversed to give the *land breeze*. The sea breeze usually begins in the forenoon and reaches its maximum a few hours after local noon. The land breeze is not as deep as the sea breeze and often more difficult to detect. *Valley winds* also are the result of heating of the ground and are important for their effect on crops. In a valley with steeply sloping sides there is a well-defined sequence, beginning before sunrise with a steady movement of air down the valley to the plain. As the sunshine becomes stronger the sides of the valley and its floor become warmer, and buoyant air moves up the slopes to descend again in the center. In the early afternoon this circulation is joined by a drift from the plain into the valley, which ultimately dominates in the late afternoon. In the evening the air in contact with the upper slopes of the valley cools and becomes more dense. This causes a "katabatic," or downslope, wind which in the late evening produces a circulation the reverse of that formed in the morning, namely, a downhill motion with ascent at the center of the valley. Late at night the flow from the valley into the plain begins and by sunrise becomes the main motion.

Installations such as nuclear reactors, which are potentially dangerous, are often sited in hilly country because of the need for an abundant supply of pure water. It is now customary to study the microclimatology of the district in order to ascertain if and where local circulations are likely to set up pockets of high concentration of effluent. Although mathematical analysis of the problem of local winds has yielded promising results, it is usually necessary to examine the motion of the air experimentally. This can be done by the use of an accurately made relief map of the district in a wind tunnel, supplemented by field studies with smoke or no-lift balloons. This application of micrometeorology is likely to become of increasing importance in the future.[4]

[4] See *Meteorology and Atomic Energy,* U.S. Weather Bureau for the U.S. Atomic Energy Commission, 1955.

CHAPTER 8

FORECASTING: OLD AND NEW

When an applied mathematician calculates the rate of flow of water in a pipe, or the strength of an electric field, or the temperature of a body which is losing heat at a known rate, or, in fact, does any of the calculations familiar in laboratory physics, he is making a forecast. In most instances the forecast is both precise and accurate—precise because it is usually expressed in numbers and accurate because it is based on the application of well-established laws of physics to relatively simple situations.

It is believed by most meteorologists that forecasts of this kind are not possible for the weather. A statement of the type "Tomorrow, at 3:42 P.M., a thunderstorm will break over Washington, D.C." is never likely to be issued by the U.S. Weather Bureau, no matter how skilled its forecasters become. Events in the atmosphere are not as closely defined as this, the very nature of thunder precludes its prediction to the minute many hours ahead, and it is never certain that convection will cause a thunderstorm. The reason for this apparent indeterminacy in atmospheric processes is not difficult to understand. Many atmospheric phenomena, in their initial stages, are below the threshold of meteorological observation, and a disturbance which is not detectable as an entity on the weather map on, say, Monday may become the dominant feature of the weather by Friday. This is particularly true of those aspects of weather that depend upon convection.

All predictions of forthcoming weather are statements of the *probability* of an event, and their reliability decreases markedly with the interval of time involved. It is possible to forecast weather with great precision and reliability a few minutes ahead simply on the basis of experience; we are accustomed from childhood to look at the sky and to say that rain is about to fall. To extend the period of the forecast from minutes to hours is rarely possible without the aid of instruments and observations from distant parts, but it is sometimes done with reasonable success by those who live in the open and have developed a keen "weather sense." Much of this experience is embodied in "weather lore," proverbial sayings that have served man in lieu of weather charts from ancient times. In the days of sail every captain had to be his own meteorologist, and some of the most reliable of the traditional proverbs are really mnemonics to aid the young sailor to learn to judge the weather by the appearance of the sky.

Weather proverbs are reliable only when they refer to the weather a short time ahead, usually not more than a few hours. The many traditional sayings that predict the nature of a coming season from the appearance of plants or the behavior of animals have no validity. It is still believed by countryfolk in England that an unusually plentiful crop of wild berries in the fall indicates that the coming winter will be severe, but accurate observations have failed to show any grounds for the tradition. Some of the ancient superstitions about weather are so unreasonable that it is difficult to imagine how they came into existence. Thus it was once held that the weather experienced on the first twelve days of the year indicates the general character of the weather in every month of the coming year—if January 5, for example, was dry, a dry May would follow. Such a belief can be described only as pure superstition. Unfortunately, meteorologists can do no better and there is as yet no method, scientific or otherwise, which can be used to forecast reliably the weather many months ahead.

In the scientific study of the weather it is customary to classify forecasts according to the period to which they refer. *Short-range forecasts* usually indicate the weather from twelve to twenty-four hours ahead, and they are often coupled with an "outlook" which is intended to be a guide for the next day or two. Forecasts of this kind are published by all the national services. *Extended-range forecasts* usually apply to periods of from four or five days to a week ahead, and are much less precise than the short-range predictions. The term "long-range forecast" is usually reserved for statements about the general character of the weather of a coming month, or possibly longer. Such forecasts are necessarily imprecise —often they rarely state more than that in named regions the period in question will be above or below average as regards rainfall and temperature—and as yet are not very reliable. Official long-range forecasts of this kind are published in some countries, notably the United States, France, Germany and the U.S.S.R., but so far many of the other major meteorological services regard the subject as still at the research stage and produce such forecasts only for experimental purposes.

Short-Range Forecasts

In the official meteorological services the daily short-range forecast is based upon the synoptic weather chart.[1] From the early days of scientific meteorology it was realized that such charts could be prepared only if an international scheme could be agreed upon for both observations and their communication, and from the middle of the nineteenth century, when the invention of the electric telegraph made communication rapid and certain, there have been numerous conferences to settle the details of the world-wide exchange of meteorological information among the nations. Unlike many international projects, the free dissemination of meteorological information has presented few difficulties of a political nature. At present all questions relating to international meteorological communications are decided by the World Meteorological Organization, a Specialized Agency of the United Nations with its headquarters at Geneva, Switzerland, and a membership of over a hundred states and territories. WMO, as it is known among meteorologists, has a small permanent secretariat drawn from all parts of the world, but the greater part of its work is done through technical commissions and regional associations, the members of which come from the national services and the universities and work on a voluntary basis. There is a long tradition of harmonious relations between the meteorological services of the world, and it is probably fair to say that there are virtually no "official secrets"

[1] See Appendix I.

in meteorology. Information concerning methods of forecasting, for example, is freely passed between the national services and it is customary for one service to help another by training its staff when needed. Weather knows no frontiers and meteorologists are among the most traveled people in the world of science. It is only in time of war that weather observations become closely guarded secrets.

The meteorologist is insatiable in his demands for more and yet more observations, and in the more settled parts of the world the network of observing stations is now very dense. In the United States there are about 400 airfield stations which supply hourly observations, but these form only a small part of the whole. All told, about 9,000 co-operative part-time stations report to the Weather Bureau, and in addition there are frequent messages from ocean weather ships, merchant vessels, Coastguard stations, and aircraft in flight. The international exchange also is on a vast scale and extremely efficient, and the forecasters in London, Paris, Moscow, Washington and other major centers know about changes in each other's weather almost before they have reached the radio and the press in the country concerned.

The observations on which forecasters depend are made at fixed hours and in a prescribed form approved by WMO on behalf of all its member states. At a full synoptic reporting station the following elements of weather are observed at fixed hours:

present weather
wind direction and speed near the ground
amount and form of cloud and height of cloud base visibility
air temperature and dew point
barometric pressure and barometric tendency
past weather (i.e., weather between the time of the present observation and the previous observation).

This information is put into the International Meteorological Code. This code consists of groups of five figures arranged in a definite sequence. Thus, for example, the chart plotter knows that the group 95854 means that the pressure at the station, corrected to sea level, was 995.8 mb, and that the temperature in the screen was 54° F at the time of the observation at the station, which is identified by another group of figures at the beginning of the message. The entire message may have been composed by a man who spoke only English, but it is immediately intelligible to someone who knows only Spanish or any other language.

These messages are collected at local centers and transmitted to national centers by teleprinter or radio. From the national centers the collected messages are retransmitted at high speed over the international network. The volume of information communicated in this way is very great; for example, the Central Forecasting Office of the United Kingdom at Dunstable, England, handles more than half a million five-figure code groups every day. In addition to the surface observations, the results of soundings by radiosondes and radar-wind equipment are also passed over the network.

At the forecasting offices the procedure varies according to the methods favored by the national service concerned, but there are certain common features. In all instances the data received in code are plotted on a weather map, again using standard symbols arranged in a prescribed order around the dot on the map which indicates the position of the station. From these data the analyst prepares the charts which show the pattern of the isobars, contours and thickness lines, and puts in the fronts. In some instances the observations are plentiful and reliable, and the meteorologist has little difficulty in constructing both surface and upper-air charts. In other cases, notably over the oceans, data are scarce and sometimes not too accurate, and much judgment has to be used. It is here that long experience tells.

When the charting and analysis are completed, that is, when the forecaster is satisfied that he knows the weather and has a fair understanding of the physical processes dominant in the atmosphere at the time of the observations, the most difficult and uncertain part of the whole process of forecasting follows. The meteorologist has to deduce from the existing chart the likely movements of the pressure patterns and their consequences during the ensuing period. Usually this is done by the preparation of a *forecast pressure chart* (sometimes called a *prebaratic*) which shows the patterns of the isobars or contour lines as the forecaster thinks they will be at the time of the next observation. In the classical method of forecasting this is done on the basis of experience coupled with some well-established rules learned from a study of past sequences of charts. Calculation, in the strict sense of the word, is not often employed. From the forecast pressure chart the forecaster writes down the expected sequence of weather, again using his experience and knowledge of the physics of the atmosphere as a guide. All this has to be done to meet a deadline, and it is a common complaint among forecasters that they rarely have enough time to weigh up a situation as thoroughly as they would like.

The classical method of short-range weather forecasting is based upon a process similar to that known in mathematics as *extrapolation,* the prediction of the behavior of a system from knowledge of its past. As a consequence, weather forecasting is highly subjective, and it is certain that two forecasters, of supposedly equal skill, working independently with the same data, would not produce completely identical forecast pressure charts and weather forecasts. In most instances the differences would not be significant, but in difficult situations two meteorologists might, with good reasons, take very different views of probable developments in the next twelve or twenty-four hours, and thus produce markedly different forecasts. For this reason most forecasts emerge as a result of a group discussion.

Even if the forecast pressure chart is substantially correct, it does not follow that the prediction of weather also will be highly accurate. The reader who has followed the accounts of weather systems given in the previous chapters will appreciate that there is no unique relation between a pressure pattern and the weather, despite inscriptions on household barometers. The accurate location of a front does not mean that the weather in that region must follow a rigid pattern, for in the event there may be heavy precipitation or simply thickening of cloud. Many of the errors that arise in forecasts are caused by incorrect timing rather than a faulty appreciation of the physical processes pertaining at the time. Fronts rarely move at a fixed foreseeable speed and they may change direction rapidly, with the result that the timing of the onset of rain can be badly in error.

Forecasts are estimates of the likelihood of certain events and are couched in language that has to be intelligible to all who use them. As a result a forecast is a literary composition, the accuracy of which is often uncertain. It is an extremely difficult matter to devise an objective method for the assessment of forecasts. There are several factors to be considered in judging the merits of a forecast. Conditions in the atmosphere do not change instantaneously and there is a certain amount of "persistence" in all meteorological elements. For instance, even in a country like Britain, where weather varies rapidly, a "forecast" that simply repeated today's weather as tomorrow's forecast would, in a long sequence, be more often right than wrong, especially with regard to temperature. Further, it is possible to make reasonable estimates of coming weather on the basis of climatological averages alone, and the task of the forecaster can be described as that of detecting impending deviations from the climatic "normals."

A "forecast" which relies entirely on climatological averages or simply uses persistence (i.e., repeats today's weather as tomorrow's forecast) may thus be right more often than it is wrong in a long sequence. Yet clearly such "forecasts" are entirely mechanical and their composition involves no knowledge of the physical processes of the atmosphere—in the language of meteorologists, no *skill* enters into their preparation. Thus, for example, temperature is one of the more persistent elements, for adjustments for inequalities of heat take place rather slowly. As a result, "persistence" forecasts of temperature are often good. If weather forecasts are to be justified on economic grounds they must display skill, that is, do better than climatological averages or persistence. The value of the synoptic-analysis method lies in the fact that it allows large and abrupt changes of temperature and other elements to be predicted and timed with tolerable accuracy. This cannot be done by purely statistical methods.

These considerations indicate some of the difficulties that make the evaluation of forecasts such a controversial topic among professional meteorologists. The main difficulty of verification lies in the heterogeneity of atmospheric phenomena. A forecast has to describe the weather of a large area in a few words and therefore cannot be absolutely correct for all places and at all times. Verification should therefore be carried out at many stations simultaneously if a meaningful result is to be obtained unless, of course, the forecast is deliberately designed to apply to one locality for a short period, for example, a landing forecast supplied to aircraft.

Many formulas have been devised to estimate the amount of skill shown in forecasts. The simplest of all is that of P. Heidke, namely

$$\text{skill} = 100\left(\frac{R-E}{T-E}\right) \text{ per cent}$$

where R is the number of correct forecasts of an entity, E is the number expected to be correct if only persistence or climatological averages were used, and T is the total number of forecasts. Thus, if all forecasts were correct ($R = T$), the skill would be 100 per cent. However, if the forecaster did no better than a machine which simply drew the forecast from the climatological records or repeated the present weather ($R = E$), the skill would be zero. Suppose that out of 20 forecasts of rain the number of successes (R) was 14, and that persistence alone (rain continuing during the forecast period) would have given correct forecasts on 9 occasions, the skill shown would be

$$100\left(\frac{14-9}{20-9}\right) = \frac{500}{11} = 45.5 \text{ per cent}$$

which is a high score. This test is undoubtedly severe and perhaps misleading, for it gives no credit to the forecaster for agreeing with the "chance" forecast, even though he may have reached his conclusions by impeccable physical reasoning.

It is impossible to give reliable figures for the "accuracy" attained in routine forecasts by the national services, for a variety of reasons. In the first place, a great deal depends on what the investigator means by "accuracy of forecasts." In some instances it is possible to be specific, for example, meaningful comparisons can be made between forecast and actual winds on specified air routes over areas with a good network of upper-air stations, but for the elements of weather the situation is much more vague. Forecasts of rain can be judged by intensity, amount, extent and duration and the published forecast may refer to none of these in detail. Secondly, forecasts often refer to a "risk" or "probability" of certain phenomena, and they are not necessarily failures if the event is not reported. From time to time verification figures are stated by the national services, but usually these refer to some specific element (such as temperature) and have other limitations, and an over-all figure of merit for general forecasts would be so vague that it could mislead.

Extended-Range Forecasts

Methods of preparing forecasts for periods of the order of a week ahead differ greatly from country to country. The common basis appears to be that if the daily pressure charts for a large area of the globe are averaged over several days the individual irregularities disappear and definite large-scale features emerge. Such features, called *centers of action,* are usually some three or four times larger than the individual circulations shown on the daily charts. The question immediately arises whether these centers of action are statistical fictions or real dynamical entities, the identity of which is hidden on the daily charts because of the many smaller circulations that are the preoccupation of the short-range forecaster. The arguments in favor of the reality of the large-scale systems turn mainly upon their persistent recurrence in the same geographical area and the fact that they show a distinguishable continuity of movement which is very similar on charts prepared with different averaging periods.

Looked at in this way, the extended-range forecasting problem becomes that of identifying mean circulation patterns from which it is possible to predict general weather characteristics and anomalies (i.e., departures from the climatological averages). The creed of the extended-range forecaster is that the broad features of weather for the next week in a certain region, whether it will be generally colder or warmer, or wetter or drier than the average, is governed by the behavior of the centers of action. He believes that if it is possible to determine in advance the trend of the large slow-moving centers of action he can obtain a first approximation to the weather that lies ahead.

The most successful exponent of this method of forecasting in recent years is undoubtedly the American meteorologist Jerome Namias, who has brought the technique to the operational stage. From Namias' work it appears that the pressure field must be studied on a hemispherical basis, because all the centers of action appear to be linked, with a fairly rapid exchange of energy between the systems via the long waves. The process of constructing the pattern of the large-scale circulation for the period that lies ahead is naturally complicated and involves the detailed study of the kinematics of the motion field, as well as the thermodynamic relations.

The level of success attained so far by this method is not high, but this is true of all methods of extended-range forecasting. Nevertheless, Namias' work forms a notable chapter in the development of meteorology and his studies of the mean-pressure charts have thrown much light on the working of the atmospheric engine.

Long-Range Forecasts

The earliest long-range forecast of which we have a record is recounted in the Bible, when Joseph attracted the attention of Pharaoh by the prediction that there would be seven years of good harvests followed by seven lean years. This is an example of the use of *cycles* or *periodicity,* and since Joseph's day meteorologists have devoted much time and energy to the search for periodicities in atmospheric phenomena.

Periodicity is familiar in astronomy, and man must have recognized certain easily distinguishable features, such as the regular waxing and waning of the Moon, at a very early stage. The periodicity of the seasons is the basis of husbandry. Astronomical periodicity means that at certain times a certain configuration of the heavenly bodies is always observed. But exact periodicities are not found in the atmosphere, and in the words of two British meteorologists, C. E. P. Brooks and N. Carruthers, "A periodicity of the kind mostly found in meteorological data postulates merely that at a certain time there is a greater or lesser probability that something will happen than at some other time; in other words, the probability is periodic rather than the event." The study of periodicities in weather is undoubtedly hampered by the lack of really long sequences of reliable observations. The meteorologist who wishes to go back for several centuries finds that he is often forced to rely upon vague historical records rather than accurate measurements. Thus the famous 35-year cycle of rainfall, put forward in 1890 by Brückner, a Swiss meteorologist, was derived from records of European weather, river heights and floods from 1691 to 1870. The Brückner cycle is not now regarded as very definite, and there are somewhat better grounds for a belief in a 52-year cycle in British rainfall, but even here the evidence is not strong. In 1925 Sir David Brunt made a detailed examination of the variations of pressure, temperature and rainfall at six widely separated European observations which had long series of reliable and coherent observations. He concluded that there is no way in which periodicities can be used with any degree of confidence for long-range forecasting. A periodicity may be quite strongly marked for some years and then disappear, and apart from annual and diurnal oscillations it seems that meteorological periodicities are never sufficiently persistent to be of use in prediction.

The solar beam is the source of all weather and it is therefore natural to look for variations in the Sun as a clue to long-period changes in terrestrial weather. Sunspots exhibit a quasi periodicity of about 11 years in their numbers, and they are easily observed. The nature of sunspots has not yet been fully established but they are undoubtedly areas of anomalous activity on the Sun's surface, and it is therefore plausible that there should be some connection between anomalous weather on the Earth and sunspot activity. Despite much labor, no relations have yet been found which are close enough to be of value in forecasting.

Many other indirect methods have been tried, so far with little success. Sir Gilbert Walker, a distinguished British mathematician, who at one time was director of the Indian Meteorological Service, spent much of his long life in a search for empirical relations between the weather in different parts of the world. His most famous investigation was into the monsoon rains of India. He developed a forecasting equation which related the departure of the monsoon rainfall from the average to elements such as the barometric pressure in South America in April and May, the rainfall in May in Zanzibar and the Seychelles, the May snowfall in the Himalayas, and the monsoon rainfall and pressure in India in the previous year, but the predictions made by this equation are not very reliable.

Methods of this kind are statistical, that is, they seek to establish relations which have a certain amount of physical plausibility without inquiring closely into the direct sequence of cause and effect. So far all such methods have failed to give really useful and reliable results. Another approach, which is still indirect, but possibly based on better ground scientifically, is that of *analogues*. It is proverbial that weather never repeats itself precisely, and as far as existing records go there is nothing to show that on any two occasions the atmosphere over a large part of the globe had exactly the same properties. On the other hand, there is a marked resemblance between situations at different times, and much of the art of forecasting depends on the meteorologist's memory of the weather which accompanied previous characteristic distributions of pressure and temperature. For parallel distributions the sequence of weather should be similar, at least for a short period ahead.

This concept has been applied to long-range forecasting by the use of *temperature-anomaly analogues*. If the average screen temperature for a period of, say, 30 days is compared with the climatic average for the same period, it is found that the anomaly, or difference between the two averages, is usually distributed in a recognizable pattern over a large area, such as the Northern Hemisphere. For example, the pattern for a given month might reveal an area of positive anomaly (warmer than average) over Canada with negative anomaly over Europe, and so on. When such a chart has been constructed, a search is made for a year in which a similar distribution of anomaly was present. It is rare to find two years with really close resemblance of temperature-anomaly distribution in a given month—usually there are several years with a vague resemblance. One of these years is picked as the year of best fit, and the long-range forecast then repeats the weather observed during the following month in that year.

The method has been tried in the Meteorological Office of the United Kingdom. The results are about the same as with other systems, that is, the over-all standard of accuracy is not high. There have been some spectacular successes, and some equally spectacular failures, in different months. There is definite evidence of skill, but not yet enough to justify the publication of the forecasts.

The long-range forecasting problem is the most intractable in the whole of meteorology and it is by no means certain that it will ever be solved. The advent of the high-speed electronic computer, however, makes feasible calculations which before could not be contemplated because of the manpower required. Meteorologists everywhere await with interest the results of these investigations, which clearly must be integrated with studies of the general circulation of the atmosphere.

Numerical Forecasting

We come now to the latest and in many ways the most promising development in the technique of weather forecasting. It has long been a dream of meteorologists to eliminate the subjective element from forecasting. The obvious way to do this is to follow the example of astronomy and to use the classical equations of motion of fluids to calculate the movements of pressure systems. There have been many attempts to obtain objectivity by the use of simplified models of atmospheric disturbances. It is possible, as a crude first approximation, to regard the typical depression as a rotating mass of air which undergoes translation at the same time (a "cartwheel vortex") and it is then a fairly easy matter to calculate the properties of the system if dissipative forces such as friction are disregarded. The result, however, is of no use for practical forecasting, for the idealization is too severe.

The modern concept of numerical forecasting was born in the fertile mind of the Quaker mathematician Lewis Fry Richardson (1881–1953) during the period of World War I. Richardson was attracted by the successes of spherical astronomy to ask why similar methods could not be developed for meteorology. In spherical astronomy the apparent motions of the heavenly bodies are calculated with great precision many years in advance, and there is never any doubt that a solar eclipse, for example, will occur exactly at the time calculated. The only doubt is whether it can be seen from the ground along the path of totality, for this depends on the presence or absence of cloud, which cannot be calculated in advance. The problems of spherical astronomy and weather are, of course, very different. Astronomy deals with relatively simple geometrical relations, whereas meteorology has to struggle with an enormously complex mechanical and thermodynamical system. A shower of rain is a far more complicated event than an eclipse.

Richardson's creed was that despite these differences there is no need to despair. He believed that it should be possible to proceed from an initial state of the atmosphere to a future state by a purely mathematical process which did not involve the use of "mechanical" models like that of the "cartwheel vortex," and he devised a method of calculation by which the process can be carried out. This method is known as *step-by-step integration* of a differential equation.

The atmosphere is a fluid, and its behavior is dictated by certain immutable laws of physics, which are summed up in the equations of motion and of thermodynamics. In theory, these equations can be solved, that is, velocity, temperature and pressure can be found by mathematical operations, provided the state of the atmosphere is known at some instant together with the boundary values, which describe the behavior of the system at the surface of the Earth and at great heights. Such solutions should describe the behavior of the atmosphere for an unlimited time ahead and, in theory, the problem of forecasting the weather can be completely solved.

During the long night of World War I, Richardson, who served with a Friends' ambulance unit in France, held fast to this vision and in 1922 there appeared one of the most remarkable books in meteorology, *Weather Prediction by Numerical Process*. In this work Richardson produced a set of equations which he thought should give a tolerable representation of the physical processes that govern atmospheric phenomena, together with a method for their approximate solution. This was the beginning of numerical forecasting. At the time it must have seemed to many that it was a forlorn hope, for the one example worked out in the book was a spectacular failure. Richardson, by immense labor, calculated the change in pressure over a period of six hours for an area in Europe and obtained an answer which was in error by at least two orders of magnitude. It must have required considerable courage to publish such a ludicrous result. We have reason to be thankful that he had such courage, for the example made clear some of the intrinsic difficulties of the problem and helped to point the way to the future.

There are at least three basic difficulties to be overcome in any purely mathematical system of weather forecasting. First, the equations of fluid motion are of the type known as "nonlinear," which means that they cannot be solved by straightforward methods, and elaborate numerical techniques have to be employed. The second difficulty is that the initial and boundary states of the atmosphere can never be known with sufficient accuracy, and in sufficient detail, to make the results comparable in accuracy with, say, the predictions of spherical astronomy. Richardson spoke of the possibility of an almanac of the atmosphere, a volume in which it would be possible to look up the weather for any future epoch, just as it is possible to find the apparent future position of the stars in an astronomical almanac. It is now common ground among meteorologists that the nature of weather is such that its exact prediction in almanac form for long periods ahead is impossible in principle. Finally, there is the fundamental difficulty that the vertical motion is small compared with the horizontal motion. The basic reason for the failure of Richardson's trial calculation was that he was compelled, with the mathematical system employed, to deduce this motion as the difference of two large quantities which were very nearly equal, and in doing so he computed a large spurious convergence.

Despite these difficulties, Richardson's work is of permanent value and the step-by-step process which he used is the basis of the modern technique. This process is simple in principle but complicated and laborious in practice unless machines are employed. Suppose that we know the barometric pressure and the rate of change of pressure (the barometric tendency) at some point at some zero of time. We can use these facts to predict the pressure a short time ahead, say one hour later, with fair confidence, but if we continued to use the same rate of change of pressure to forecast the pressure a long time ahead, say twelve hours, we should undoubtedly be in danger of serious error. Because of the rapidly changing pressure pattern the barometric tendency measured at a point on the Earth can be trusted to retain its original value for short periods only, and the only safe method is to repeat the process over and over again with new values of the rate of change as they become available. The step-by-step process of solution consists of a series of short-range forecasts which ultimately cover a long period of time. It is thus laborious and excessively tedious for the human brain, but admirably adapted for use with a machine such as a high-speed electronic digital computer.

A digital computer consists of an *arithmetical unit* which performs elementary operations such as addition and multiplication, a *store* (or *memory*), and a *control system.* The characteristic feature of the modern electronic digital computer is the prodigious speed at which the elementary operations are carried out. This feature makes it possible to replace the exact differential equations by approximate algebraic equations which can be solved by arithmetical methods by the machine in a short time. To do this the computer must perform certain operations in the correct sequence; this is called the *program* and the instructions are usually fed into the machine by punched or magnetic tape. The *data,* that is, the initial meteorological observations, are also fed in by tape, and the computer proceeds to solve the equations over and over again at hundreds of points over the area for which the forecast is required. In this way, starting from the observed pressure distribution at the zero of time, the machine proceeds to construct the future pressure pattern for 12 or 24 hours ahead by steps corresponding to half-hourly or hourly intervals of real time. A modern computer can perform this feat, which involves millions of mathematical operations, in a few tens of minutes. The final result is a forecast pressure chart which is printed by an electric typewriter in a form which the forecaster can use to deduce the coming weather.

At the time that Richardson wrote his book the electronic digital computer had not been invented, and he was forced to use a desk machine for his trial computation. With such machines he estimated that to keep pace with the weather of the world would require no less than 64,000 mathematicians working together. Numerical forecasting, as a practicable scheme, could not have been achieved without the development of high-speed electronic computing devices.

Forecasting by Equations

The differential equations that describe the motion of the atmosphere were not discovered by Richardson; they had been known since the eighteenth century. Their use in the problem of weather is not straightforward, and numerical forecasting is not simply a matter of feeding the data and the original equations into a high-speed computer and waiting for the result to appear. The modern approach to the problem, which is largely the joint work of the late John von Neumann and Jules Charney, makes considerable use of physical concepts which have emerged as the result of the studies of the synopticians.

The dynamical systems that are the main concern of the short-range forecaster have, as we have seen, certain characteristic properties. Their vertical motion is small compared with the horizontal motion, their vertical accelerations are small compared with gravity, and their horizontal accelerations are small compared with that which appears in the Coriolis effect. Finally, when examined over periods of the order of 12 or 24 hours, the thermodynamic processes involved are approximately adiabatic, that is, changes in temperature are the result mainly of changes in pressure.

These considerations suggest that reasonable approximations to the behavior of the weather-producing systems can be obtained if the systems are considered to be two-dimensional, with pressure and height related by the hydrostatic equation, and without friction. When temperature depends solely upon pressure, the fluid is said to be *barotropic*.[2] In such a fluid there is no change of wind with height, and because of the absence of friction and of non-adiabatic changes of temperature (such as the air becoming warmer by passing over heated land) it is to be expected that the barotropic model is most likely to be successful in the upper air. Experience has shown that numerical forecasts which depend upon this simplified system of fluid dynamics are at their best at the level of nondivergence, the 500 mb surface which lies at about 18,000 feet above sea level.

The equations used are those of a frictionless fluid which moves

[2] See Appendix I.

horizontally subject to the geostrophic balance. For the purpose of the calculation they are arranged to bring out the influence of vorticity, and in particular of the conservation of absolute vorticity (see Chapter 4). The problem may then be stated in the simplest possible terms as that of finding how the pressure pattern changes from a given initial distribution when vorticity is transferred by the wind but always so that its absolute value is conserved. The starting point is a set of contour heights of the 500 mb surface over a large area, and the final result is a new set of contour heights which shows the shape of the same surface at the end of the forecast period. If the equations expressed all the physical processes operative during the period, and the approximations introduced no spurious effects, the result would be a perfect forecast of the pressure pattern at about 18,000 feet some 12 or 24 hours ahead. It is important to realize that a forecast of weather is not attempted, and that the scheme reveals nothing about features such as clouds.

The practical method adopted is as follows: A large rectangular area, such as that covering the United States and parts of the adjacent oceans, is divided into several hundred rectangles of side about 300 kilometers and numbered as in Fig. 48. Radiosonde ascents give the values of the height of the 500 mb surface (h) at many points in this area. From these observations the values of h at the lattice points of the mesh are found by interpolation. (This task is now often done by a machine, a process known as *objective analysis.*) The data on which the computer has to work are thus values of h at a selected zero of time at every lattice point of the area.

The mathematical basis of this work is outlined in Appendix II, together with a brief account of the method of solution. The process naturally involves advanced mathematics, but the nonmathematical reader should be able to follow the main ideas from the following simplified account.

The starting point is the hydrodynamical equation which expresses the principle of the conservation of absolute vorticity,[3] that is, no matter how air moves over the surface of the Earth it keeps its absolute vorticity unchanged. This equation has to be related

[3] See Chapter 4 and Appendix II.

to changes in the height of a surface of constant pressure (usually, the 500 mb surface), and for this purpose it is assumed that the pressures and winds are in geostrophic balance. This assumption enables the velocity and the vorticity of the winds to be expressed mathematically in terms of the slopes—the ups and downs—of the constant-pressure surfaces. This is equivalent to expressing these properties in terms of pressure gradients but, as explained before, it is more convenient for the mathematician to use the height of the

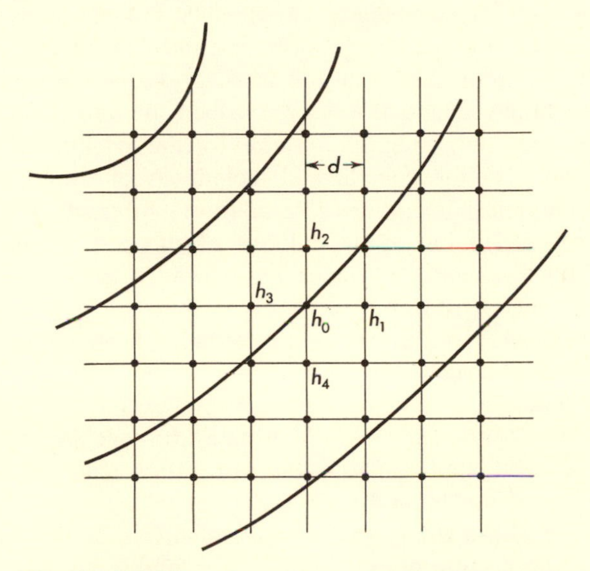

Fig. 48. Division of area for numerical forecasting.

constant-pressure surface rather than the actual pressure. After some manipulation, the equation becomes one in which the time rate of change of contour height (really, the barometric tendency) is the unknown quantity. This is the *forecasting equation,* which is now ready to be solved at all the lattice points by the machine.

The forecasting equation is what is called a *second-order partial differential equation* and is of the kind known to mathematicians by the name of Poisson. Equations of this kind are familiar in mathematical physics, especially in the theory of electricity. In many instances a Poisson equation can be solved exactly, that is, the unknown quantity can be expressed by a single mathematical formula. This is not possible in the meteorological problem, and approximate numerical methods have to be employed. There is no magic formula to be discovered which will predict the weather during the next 24 hours, and the answer provided by the machine is simply a long list of figures.

High-speed computers are sometimes called *electronic brains,* but this is misleading. Machines of this type cannot reproduce the subtle and ingenious methods by which a skilled mathematician solves a partial differential equation, and in this respect a computing machine is more like a moron who can be put to work indefinitely on simple repetitive tasks prescribed for him by a superior intelligence. In composing the program for the machine the mathematician replaces the complicated techniques of the calculus by simple arithmetical operations like addition, subtraction and multiplication, carried out many millions of times. But approximations of this kind must introduce errors, which are often difficult to reduce. Such "truncation errors," as they are called, may be partly responsible for certain characteristic shortcomings in numerical forecasts, such as a tendency to overestimate the intensity of anticyclones.

With the program, and the initial data, the machine can be set to work to solve approximately the forecasting equation at every one of the grid points, and thus to determine the rate of rise and fall of the height of the constant-pressure surface as the air moves subject to the constraints imposed by the physical laws adopted, in this instance chiefly the law of the conservation of absolute vorticity. The rates of change (or barometric tendencies) so calculated are sufficiently accurate to allow the machine to make a forecast of pressure values at, say, the end of the next half hour of natural time. The result is a new set of contour heights and thus a new shape for the 500 mb surface—a hollow, or low, may have deepened or filled in a little or a high intensified. The new heights are then passed to the store or memory of the computer for use in the next stage of the calculation, when the solving process is repeated with the new heights as the initial data, and so on until the whole of the forecast period has been covered in half-hourly steps of natural time. The actual time taken by a modern machine to complete a half-hourly step is less than a minute.

A numerical forecast of pressure changes in the next 12 hours thus involves the solution of the forecasting equation and the prediction of the pressure change 24 times in succession at hundreds of points, if half-hourly steps are employed. If the steps are shortened to, say, 15 minutes of natural time, the approximation errors are likely to be somewhat reduced, but the number of operations, already running into millions, is increased. Numerical forecasting, even with simple models, calls for very fast machines, for nothing is more useless than a late forecast.

The barotropic-fluid scheme described above is sometimes called the *one-parameter model,* for motion at one level is characteristic of motion at all levels. It has been used with success to forecast winds at 20,000 feet, where the predictions of the mathematician are comparable with, and sometimes better than, those of the "conventional" forecaster. However, a forecast for one level cannot be satisfactory at all levels, and it is necessary to take into account thermal effects. The real atmosphere is not barotropic but *baroclinic,* that is, large horizontal gradients of temperature are present, giving rise to thermal winds. In terms of numerical forecasting this means that not only contour height but thickness must be brought into the calculations. Mathematical schemes which allow for some of the effects of baroclinicity have been devised; the results are not notably more accurate than the simple barotropic models at great heights but they usually show some improvement for the surface layers. The simple models also make no allowance for the effects of topography, whereas every forecaster knows that great mountain ranges such as the Rockies have an appreciable influence on the movement of pressure systems. Schemes which remove the "flat earth" limitation are now the subject of intensive research. Much work is also being done to allow for the effects of nonadiabatic heating and of friction, with the object of improving forecasts at low levels.

It is still too early to attempt an objective evaluation of the impact of numerical forecasting upon meteorology. In the more advanced models it is possible to estimate the all-important vertical motion and hence to predict the intensity of rainfall, but as yet no method has been devised which allows quantitative assessments to be made of cloud and visibility, factors which are of the greatest importance in aviation forecasts. In broad terms, the human brain is still needed to complete the machine-made forecast by "putting in the weather," and this situation is not likely to be changed in the near future.

The introduction of genuine mathematical methods into forecasting is a major advance in meteorology, comparable with that achieved by the Norwegians when they brought air-mass analysis and the notion of discontinuities as significant features of weather systems into the highly empirical art of weather prediction. It is the biggest advance yet made in the progress of meteorology toward the status of an exact science and Richardson, who died in 1953, must have rejoiced to see his dream realized, if only in part.

APPENDIX I

A SHORT GLOSSARY OF TECHNICAL TERMS

ABSOLUTE TEMPERATURE Many different scales, notably those of Celsius (centigrade) and Fahrenheit, have been used for the measurement of temperature since the invention of the thermometer. These scales use certain easily reproducible temperatures, such as those of melting ice and boiling water, as fixed points. In the science of thermodynamics, and in many calculations, it is necessary to employ a scale which is independent of the properties of a specific substance, such as water. Such a scale is the Kelvin or absolute scale which, for most practical purposes, is simply the centigrade scale with 273 degrees added. In the Kelvin scale the average temperature of the air near the surface of the Earth is 288°, that is, 15° C. Temperatures may also be expressed in an absolute scale of Fahrenheit degrees by the addition of 459° F.

ADIABATIC LAPSE RATE This term is used in meteorology to denote the fixed rate at which temperature must decrease with height in the atmosphere so that a volume of air moving vertically without gain or loss of heat should always have the same density as the surrounding air. The value of the adiabatic lapse rate for dry air is about 1° C per 100 meters. See p. 34.

ADIABATIC PROCESS A change in pressure and temperature (and therefore in density) of a substance is said to occur adiabatically if no heat enters or leaves the substance during the process. If a volume of air at 1,000 millibars pressure and 17° C temperature were suddenly expanded so that its pressure fell to 900 millibars, its temperature would fall simultaneously to about 8° C. See p. 34.

ADVECTION A term used in meteorology (and rarely, if ever, in other sciences) to denote the process of transfer by horizontal motion, for example, the transfer of heat from low to high latitudes by the winds that make up the general circulation.

ALBEDO The fraction of the incoming radiation which is diffusely reflected. Some typical values are: fresh snow, 0.7 to 0.9; fields and woods, 0.02 to 0.15; whole Earth, including clouds, 0.34 to 0.45. See p. 9.

ANEMOMETER An instrument for measuring wind speed.

ANOMALY The difference between the value of a meteorological element, such as temperature, at a given time and place and its average or climatic value at the same place.

BAR The unit of atmospheric pressure. Meteorologists are almost alone among scientists in their regular use of this absolute unit (that is, a unit which does not depend upon the local value of gravity) for pressure. In less enlightened sciences the length of a column of mercury, measured in inches or centimeters, is still employed for this purpose. The bar, equal to 1,000 millibars (mb), is a million dynes per square centimeter and equals 29.53 inches, or 750.08 centimeters, of mercury at 273° K (32° F) when gravity has the standard value. See DYNE.

BAROCLINIC, BAROTROPIC A baroclinic atmosphere is one in which surfaces of equal density intersect surfaces of equal pressure. This means that temperature changes in the horizontal as well as in the vertical. When surfaces of equal density and of equal pressure coincide, the atmosphere is said to be barotropic. In meteorological literature, a region of "strong baroclinicity" is one in which there are marked horizontal changes of temperature, and the thermal wind (*q.v.*) is much in evidence.

BUYS BALLOT'S LAW This law expresses the effect of the rotation of the Earth on the movement of air around centers of low or high pressure. It holds only in the extratropical regions and is usually expressed in the form "If you stand with your back to the wind in the Northern Hemisphere, the low pressure will be on your left hand," that is, north of the equator winds blow anticlockwise around a low and clockwise around a high. The rule is reversed in the Southern Hemisphere. See p. 38.

CALORIE The calorie is the unit in which quantity of heat (energy) is measured. The small or gram calorie is defined to be the amount of heat required to raise the temperature of 1 gram of pure water from 15° C to 16° C.

$$1 \text{ calorie} = 4.18 \text{ joules} = 4.18 \times 10^7 \text{ ergs} = 4.18 \text{ watt seconds.}$$

CONDUCTION OF HEAT The process by which heat is transferred from one region of a solid, liquid or gas as a result of the random motion of the molecules.

CONTOURS In meteorology, lines on a weather chart passing through places at which a constant-pressure surface has the same height above sea level. CONTOUR HEIGHTS are used in modern synoptic meteorology to display the horizontal distribution of pressure at different heights. See p. 30.

CONVECTION OF HEAT The transfer of heat by the bulk motion of a liquid or gas. If the motion occurs solely because of density differences (for example, by hot air rising from a fire) the process is called NATURAL CONVECTION If the transfer takes place as a result of motion arising from other causes, such as cold air being blown over a warm body by a fan, the process is called FORCED CONVECTION.

CONVERGENCE A term used in meteorology to describe the horizontal motion of the air into a central region. See p. 81.

CORIOLIS FORCE A fictitious force used in dynamical meteorology to simplify calculations in which the effect of the rotation of the Earth has to be taken into account, for example, the geostrophic wind equation. See p. 39.

COULOMB The unit used to measure the quantity of electricity. It is defined to be the amount transferred by a current of 1 ampere flowing for 1 second.

CYCLONE In general usage, a cyclone is a dynamical system characterized by rotation of the air around a center of low pressure. In the meteorology of the temperature latitudes, the word has the same meaning as LOW or DEPRESSION, but in other parts of the world, notably the Indian Ocean, a cyclone is a severe tropical storm or HURRICANE.

DENSITY The mass of a unit volume of a substance, measured in grams per cubic centimeter or pounds per cubic foot. The density of clean dry air at the pressure of 1,000 mb and temperature 290° K (17° C) is 0.001201 grams per cubic centimeter. The formula for calculating the density of moist air at the pressure of p mb and temperature T° K is

$$0.001201 \left(\frac{p - \frac{3}{8}e}{1000}\right)\frac{290}{T}$$

where e is the vapor pressure of the water in the air in millibars. This shows that density increases with pressure and decreases with temperature, and also that if the proportion of water vapor is reduced, the air pressure and the temperature remaining the same, the density increases. Thus in general wet air weighs less than dry air. For most purposes in meteorology the relation between the pressure (p), density (ρ) and absolute temperature (T) of a volume of the atmosphere is expressed with adequate accuracy by the *equation of state* of an ideal gas, namely

$$p = R\rho T$$

where R, the gas constant for dry air, has the value 2870.3 when p is in millibars, ρ in grams per cubic centimeter, and T in degrees Kelvin.

DIFFUSIVITY This word is used in two ways in meteorology: first, in a general literary sense to denote the ability of the atmosphere to spread suspended matter (e.g., "the diffusivity of the lower layers of the atmosphere is at its highest level in conditions of strong sunshine and high wind"), and second, as the name of a physical quantity, usually measured in square centimeters per second, which occurs in the mathematical treatment of diffusion.

DINES COMPENSATION A falling (or rising) barometer indicates that air is being lost (or gained) over an area. Yet air flows into the lower levels of a cyclone (low) and out of the base of an anticyclone (high). To explain these facts the British meteorologist W. H. Dines suggested that in a column of air there are alternate "boxes" of inflow (convergence) and outflow (divergence). In a cyclone, the high-level divergence exceeds the low-level inflow. This concept has been accepted generally and is known as the "Dines compensation." See p. 83.

DIURNAL CHANGE The variation in a meteorological element during a period of 24 hours. The obvious example is temperature, but atmospheric pressure also exhibits a regular change during a day, especially in the tropics, with maxima at 10 A.M. and 10 P.M., and minima at 4 A.M. and 4 P.M.

DIVERGENCE In meteorology, motion of the air away from a center. In mathematics, a scalar (nondirectional) quantity derived from the components of a vector (directed) quantity. See Appendix II and p. 81.

DYNE The metric system unit of force, that which causes a mass of 1 gram to accelerate 1 centimeter per second per second. It corresponds to the poundal in the imperial system.

EDDY A word often used somewhat loosely in fluid mechanics, either as a noun or as an adjective. It may indicate a volume of fluid, of unspecified size, which is forced out of its environment by influences such as the roughness of the surface or buoyancy, and thus causes random fluctuations in the flow, and ultimately enhanced diffusion. Equally, it may be used to indicate the nature of the motion, as in "eddy flow," which is a synonym for turbulence, or to indicate the primary cause of some effect, for example, "eddy diffusivity."

ENERGY The energy of a system is "its capacity for doing work." The main forms are: potential energy, which depends on the configuration of the system; kinetic energy, which is related to the motion of the system; heat energy, which is related to temperature; as well as certain other forms, such as sound, electrical, magnetic and chemical energy. All these forms are transformable into each other, and the principle of conservation of energy says that in a transformation energy can neither be created nor destroyed. In the atmosphere, potential energy is shown by the distribution of air density at various levels, kinetic energy by the winds and heat energy by changes in temperature caused by solar radiation.

ERG The unit of energy in the metric system. The practical unit is the joule (*q.v.*).

FRONT The name introduced into meteorology by Bjerknes and his associates to indicate the boundary zone between masses of air, or currents, with marked differences in temperature and humidity. In such zones the interactions between the air masses produce most strongly the characteristic features of weather.

GEOSTROPHIC WIND, GEOSTROPHIC BALANCE In large-scale extratropical atmospheric systems, the air moves so as to maintain a balance between the pressure-gradient and Coriolis (*q.v.*) forces. The motion which satisfies this balance is the geostrophic wind, which blows along the isobars. Its value in meteorology lies in the fact that the geostrophic wind approximates closely to the actual wind at heights at which the frictional drag of the surface of the earth is not appreciable.

GRADIENT In meteorology, the rate of change of an element (such as pressure, temperature or velocity) with distance. Pressure gradient is the rate of change of pressure per unit distance measured perpendicular to the isobars. Temperature gradient may be measured in the vertical, when it is usually called the lapse rate, or in the horizontal. The gradient of wind speed is usually called a shear.

ISALLOBARS Lines drawn on a weather chart through places at which the same changes of pressure have occurred in the same period. An isallobaric low is a region in which pressure has fallen, and an isallobaric high a region in which pressure has risen, more rapidly than elsewhere.

ISOBARS Lines drawn on a weather chart through places at which atmospheric pressure has the same value at the same time.

ISOTHERMAL STATE, ISOTHERMS A fluid is said to be in an isothermal state when it has the same temperature at all points. Isotherms are lines drawn on a weather chart through points at which the air temperature had the same value at a given time.

JOULE The practical unit of energy in the metric system, equal to ten million (10^7) ergs. See also CALORIE.

KILOWATT HOUR A unit of electrical energy, equal to 3.6×10^6 ergs or 8.6×10^5 calories.

LAMINAR FLOW When air (or any other fluid) moves "smoothly," i.e., the velocity does not show the large random fluctuations characteristic of turbulence (*q.v.*), it is said to be in laminar motion. The name arises because the fluid particles can be imagined to glide over each other in parallel sheets without mixing, except that caused by the infinitesimal molecular movements.

LATENT HEAT When ice melts, or water evaporates into vapor, a certain minimum amount of energy is required to bring about the change without raising the temperature of the substance. The latent heat of fusion is the quantity of heat required to change 1 gram of a substance from the solid to the liquid state without change of temperature, and the latent heat of vaporization is defined in the same way for the change from the liquid to the gaseous state. When the reverse process occurs and a vapor condenses or a liquid freezes, latent heat is released. The latent heats of water are unusually large and this is the main reason why water, which is continually being evaporated (heat absorbed) or condensed (heat released) in the atmosphere, is so important in the physics of weather systems.

MEAN This term usually implies the arithmetic mean, or the average formed by adding a number (n) of values of a variable quantity and dividing by n.

MONSOON This name is now given to systems of winds that blow regularly in different seasons of the year, but with an alternation in direction from one season to another. In eastern and southeastern Asia, the rainfall associated with the summer monsoon is an important factor in the agriculture of many lands.

PRECIPITATION In meteorology, a deposit of water from the atmosphere in the solid or liquid form (e.g., snow, hail, rain).

PRESSURE, DYNAMIC The force per unit area which arises from the motion of a fluid, for example, when a stream is brought to rest and diverted by an obstacle. The dynamic pressure of the natural wind is usually small compared with the static pressure of the atmosphere.

PRESSURE, STATIC The force per unit area exerted on a surface by a fluid at rest. The magnitude of the force is independent of the orientation of the surface, that is, pressure in a fluid at rest is the same in all directions. The static pressure of the atmosphere, the quantity measured by the barometer, is produced by the weight of the overlying air and thus decreases with height.

SCREEN TEMPERATURE At meteorological observatories, the temperature of the air is measured inside a specially designed louvered box ("Stevenson screen") placed about 4 feet above ground, so that the bulb of the thermometer is screened from direct sunshine. For this reason, screen temperatures are often popularly called "shade temperatures."

SPECIFIC HEAT Some substances heat up more easily than others. More precisely, the amount of heat required to raise the temperature of a given mass by a fixed number of degrees varies from substance to substance. In physics, this property is expressed by the specific heat of a substance, defined as the quantity of heat required to raise the temperature of 1 gram of the substance through 1° C. For gases such as air there are two specific heats; one (c_p) measured at constant pressure and the other (c_v) measured at constant volume; c_p is greater than c_v, for with a gas at constant pressure some of the heat energy is used in expansion.

SQUALL A strong wind that rises and dies away suddenly but lasts for several minutes. It is thus distinguishable from a *gust,* which is the name given to a very transient increase of the speed of the air above its average value. Squalls are often accompanied by heavy rain or hail, together with a sudden drop in temperature and a large change in the direction of the wind.

STRATOSPHERE The name originally given to the external layer of the atmosphere at the time when it was thought that above a certain level, called the tropopause, temperature ceased to decrease with height and there was no convection. The name is now used mainly for the lower part of the external layer in which the change of temperature with height is very much less than that found in the lower layer, or troposphere, which is also the region of weather. See p. 25.

SUBSIDENCE A prolonged slow downward motion of the air over a large area, as in an anticyclone or high.

SUPERCOOLING A substance, such as water, exists in three phases, solid, liquid and gaseous. Changes from one phase to another usually occur at definite temperatures—thus pure ice melts at 0° C (32° F) and pure water boils at 100° C (212° F) at standard atmospheric pressure. However, if pure water is cooled slowly and continuously, its temperature can be reduced below 0° C without freezing taking place. In this state the water is said to be supercooled. The supercooled state is unstable, and freezing begins and proceeds very rapidly if only a small quantity of ice is introduced. Other substances, known as freezing nuclei, also prevent supercooling. See p. 63.

SYNOPTIC CHART A chart which shows weather conditions over a large area at a given time.

THERMAL AND THERMOMETRIC CONDUCTIVITIES The physical quantity which measures the ability of a substance to conduct heat is called its thermal conductivity. When this is divided by the density of the substance, the result is called the thermometric conductivity, and this is the quantity that appears most often in the mathematical treatment. The thermometric conductivity is measured in square centimeters per second, and its value for air near the surface of the Earth is about 0.18.

THERMAL WIND The distribution of pressure at ground level often differs markedly from that in the upper air, mainly because of changes in the horizontal distribution of temperature in the intervening layer of air. In consequence, high-level winds often differ considerably in speed and direction from surface winds. To account for this, meteorologists have brought in the concept of the thermal wind, defined as the difference (taking into account both speed and direction) of the geostrophic winds (*q.v.*) on two constant-pressure surfaces. Thermal winds blow along the thickness lines (*q.v.*). See also p. 41.

THERMODYNAMICS The branch of mathematical physics that deals with changes in the distribution of heat and other forms of energy in a physical system. It is not concerned with the actual mechanism by which the changes are brought about and its results are therefore very general.

THICKNESS In meteorology, thickness is defined to be the difference in height between two constant-pressure surfaces, such as the 1,000 and 500 millibar surfaces. Thickness lines are drawn on a weather chart through places at which the thickness has the same value. Thickness is proportional to the mean temperature of the layer of air between the two constant-pressure surfaces, and thickness lines are thus parallel to the isotherms of mean temperature.

TURBULENCE The state of fluid motion in which the velocity exhibits large and apparently random fluctuations. See p. 130.

VELOCITY PROFILE The name given in fluid mechanics to the graph that shows how velocity varies in a definite direction (generally vertically or horizontally) in a current. The slope of the velocity profile is called the velocity gradient or shear.

VISCOSITY The property of a fluid that characterizes its power to resist deformation. A very viscous liquid is one which is difficult to pour rapidly, such as tar. In physics, viscosity is measured by two coefficients: (1) the dynamic viscosity in units of grams per centimeter-second and (2) the kinematic viscosity, which is the dynamic viscosity divided by the density of the fluid, measured in square centimeters per second. The dynamic viscosity of a gas is much less than that of a liquid, but the kinematic viscosity of air is greater than that of water, because air is much less dense than water. The kinematic viscosity of air near sea level is about 0.15 square centimeters per second. An analogous quantity, the eddy viscosity, is used to describe the properties of a turbulent fluid, but this is not a true physical constant. See p. 137.

VOLT The unit of potential difference in electricity and of electromotive force (emf) in an electrical circuit (analogous to pressure in a hydraulic circuit). A potential difference of 1 volt across a resistance of 1 ohm causes a current of 1 ampere to flow.

VORTICITY The mathematical quantity that measures rotation in a fluid. See Appendix II.

WATT The metric system unit of power, equal to 1 joule per second. In electricity, the power in watts is given by the product of volts and amperes. In the imperial system the unit is the horsepower, equal to 745.7 watts.

APPENDIX II

THE MATHEMATICAL BACKGROUND

For the benefit of those readers who wish to identify the mathematical relations referred to in the text, a brief list of the more important formulas and equations is appended.

THE HYDROSTATIC EQUATION If p is static pressure, ρ is density and z height above sea level, the fundamental static relation is

$$\frac{dp}{dz} = -g\rho$$

where g is the acceleration due to gravity.

DIVERGENCE If u, v and w are the components of velocity $\mathbf{V}$ along axes Ox, Oy and Oz, the divergence of $\mathbf{V}$ is defined to be

$$\operatorname{div} \mathbf{V} = \frac{\partial u}{\partial x} + \frac{\partial v}{\partial y} + \frac{\partial w}{\partial z}$$

The *horizontal divergence* is the above expression without the last term, i.e.,

$$\operatorname{div}_H \mathbf{V} = \frac{\partial u}{\partial x} + \frac{\partial v}{\partial y}$$

THE TENDENCY EQUATION If p_0 is the air pressure at sea level, the barometric tendency, or rate of change of pressure with time at sea level, is given by

$$\frac{\partial p_0}{\partial t} = g \int_0^{\infty} \operatorname{div}_H \rho \mathbf{V} \, dz$$

VORTICITY With the notation used above, the components of vorticity Ω are

$$\Omega_x = \frac{\partial w}{\partial y} - \frac{\partial v}{\partial z}, \quad \Omega_y = \frac{\partial u}{\partial z} - \frac{\partial w}{\partial x}, \quad \Omega_z = \frac{\partial v}{\partial x} - \frac{\partial u}{\partial y}$$

In problems of large-scale weather systems, *Ox* and *Oy* are horizontal and *Oz* is vertical. Only the vertical component of vorticity, Ω_z, is significant. If the axes are fixed in the Earth, Ω_z is called the *relative vorticity*. The *absolute vorticity* η is the relative vorticity plus twice the vertical component of the Earth's spin, i.e.,

$$\eta = \Omega_z + 2\omega \sin \phi$$

where ω is the angular velocity of the Earth about its north-south axis and ϕ is latitude.

NUMERICAL FORECASTING The following is the simplest form of numerical forecasting (the so-called barotropic model).

The model assumes that the motion is entirely horizontal ($w = 0$) and that absolute vorticity is conserved, i.e.,

$$\frac{\partial \eta}{\partial t} + u \frac{\partial \eta}{\partial x} + v \frac{\partial \eta}{\partial y} = 0$$

If h is contour height and the winds are geostrophic, the following relations hold:

$$u = -(g/\lambda)\frac{\partial h}{\partial y}, \qquad v = (g/\lambda)\frac{\partial h}{\partial x}$$

$$\eta = (g/\lambda)\left(\frac{\partial^2 h}{\partial x^2} + \frac{\partial^2 h}{\partial y^2}\right) + \lambda = (g/\lambda)\nabla^2(h) + \lambda$$

where $\lambda = 2\omega \sin \phi$ and $\nabla^2 = \partial^2/\partial x^2 + \partial^2/\partial y^2$

It is then easily deduced that

$$\nabla^2 \left(\frac{\partial h}{\partial t}\right) = -u\frac{\partial \eta}{\partial x} - v\frac{\partial \eta}{\partial y}$$

This is the *forecasting equation*, in which the unknown quantity is $\partial h/\partial t$, which is effectively the barometric tendency on the constant-pressure surface defined by h.

To solve this equation over the selected area requires (1) that h be known at all the grid points at $t = 0$ and (2) that $\partial h/\partial t$ be known on the boundary of the area for all $t \geqslant 0$. The first requirement is met from the meteorological observations at the zero of time, but for the second an estimate of $\partial h/\partial t$ on the boundary must be made in advance. Errors in this estimate will not seriously affect the forecast at points near the center of the rectangle provided that the period of the forecast is not too long.

The method of solution is based upon the replacement of the derivatives by the ratios of finite differences. In this way $\partial h/\partial t$ is evaluated at $t = 0$ at all the grid points. If δt is a small interval of time (say, half an hour), new values of the contour heights h are obtained by the use of an extrapolation formula of the type

$$h(t_0 + \delta t) \simeq h(t_0) + \left(\frac{\partial h}{\partial t}\right)_0 \delta t$$

where the suffix zero denotes values at the beginning of the forecast period. This process is then repeated with successive new values of h as the initial data until the whole of the forecast period is covered.

INDEX

About the Author

SIR GRAHAM SUTTON is the head of the British official weather service, the Meteorological Office. He was born in Monmouthshire, on the border of England and Wales, and was educated mainly at the University College of Wales, Aberystwyth and Jesus College, Oxford (of which he is now an honorary Fellow). As a scientist he has worked mainly in British Government establishments, covering a wide field which includes meteorology, chemical defense, rocket ballistics, tank armaments and radar. From 1947 to 1953 he was Professor of Mathematical Physics and later Dean of the Royal Military College of Science. In 1948 he came to America to give a course of lectures to the U.S. Weather Bureau on micrometeorology, and he has since paid many visits for scientific conferences. His previous books are: *Micrometeorology,* the standard text on this branch of the subject; *The Science of Flight,* a popular account of aerodynamics; and *Mathematics in Action,* an account of applied mathematics.

He was elected a Fellow of the Royal Society in 1949 and was knighted in 1955. His wife was a fellow student at Aberystwyth and they have two sons. They live near Ascot, England.

Set in Times Roman
Format by Jacqueline Wilsdon
Manufactured by The Haddon Craftsmen, Inc.
Published by Harper & Brothers, New York